T4-APU-517

Contents

CHAPTER

I

Introduction and Historical Background

The subject of this monograph is diagnostic cytopathology of the lower respiratory tract. It is the culmination of close collaboration for a number of years between the authors, who have shared from the beginning a great enthusiasm for the place that cytologic techniques and principles should occupy in the evaluation of pulmonary disease. One of the great advantages of this shared experience has been the opportunity for each of the authors to broaden his own base of cytologic knowledge through exposure to the material of the other. During the first eight years of the 1970s the cytopathology laboratories of Duke University and The Medical College of Virginia have examined over fifty thousand specimens from the lower respiratory tract, a number exceeded only by that recorded for examination of smears from the female genital tract. These statistics are a true reflection of the frequency with which cytologic diagnostic techniques of the lower respiratory tract are being utilized in hospitals and clinics throughout the country. In this book, then, not one, but two points of view may be expressed on many subjects. After a brief historical review, relevant cytopreparatory techniques will be discussed. Following in order will be discussions of the cytopathology of benign diseases, with a special emphasis on infectious diseases, the cytopathology of primary and metastatic cancer, early lung cancer, and transthoracic thin-needle aspiration biopsy.

HISTORICAL BACKGROUND

Perhaps it would be more than mildly surprising to many to learn that microscopic descriptions of cellular preparations from human subjects had been reported for more than a century before the early published observations of Papanicolaou[1] and Babes.[2] Exfoliated epithelial cells were described by Fontana in 1767.[3] The year 1838 witnessed not only the publication of their cell theory by Schwann and Schleiden,[4-6] but also the publication of two additional studies on the microscopic examination of exfoliated cells. The first was by Donné on fresh smears prepared from human colostrum.[7] The second was a book by Mueller in which considerable discussion was devoted to the microscopy of cancer cells.[8, 9] With these studies there began a flurry

1

of investigative activity in microscopy of body fluids and exudates, with secretions from the tracheobronchial tree in particular arousing much interest.

In 1845 Donné published the first work dealing with exfoliated cells of the respiratory tract.[7] Walshe in London in 1845 made note of the presence of tissue fragments of malignant tumor in sputum.[10-12] Beale in 1860 demonstrated malignant cells in the sputum from a patient with cancer of the pharynx.[12-14] The work of Hampeln did much to further strengthen the validity of cytologic diagnosis. In 1887 he published the report of a case in which cancer cells were correctly recognized in the sputum five months before the patient's death. The origin of these cells in a bronchogenic cancer was confirmed at autopsy. In 1919 Hampeln was the first to publish a series of cases in which sputum cytology was utilized for the primary diagnosis of lung cancer. In this series cytology was positive for tumor cells in 13 of 25 cases of lung cancer.[8, 10, 11, 15]

Thus, as the end of the nineteenth century approached, significant work already had been done which should have firmly established diagnostic cytology as a useful and important tool in modern medicine. Unfortunately, however, that was not to be the case. The utilization of this methodology greatly declined in the early part of the twentieth century and, in spite of sporadic publications, was not to be revived until the landmark publication of Papanicolaou and Traut in 1943.[16] Two major causes for this decline have been cited. The development of methods for culturing micro-organisms and the invention of the mechanical microtome diluted the interest in direct microscopic examination of smears.[17, 18] Also, some late nineteenth century medical thought questioned the morphologic identity of cancer cells as contrasted with benign ones.

Although contributions to the exfoliative cytologic literature were sparse during the first 40 years of the twentieth century, two papers merit attention here, as they jointly contributed to a more solid technical foundation. In 1928 Papanicolaou reported on a simple method of fixing vaginal smears with which fine cytologic detail could be permanently retained.[1] Similarly, Dudgeon and Wrigley in 1935 successfully applied a wet-film fixation in a mixture of ethyl alcohol, mercuric chloride, and acetic acid to the examination of sputum for neoplastic cells.[19]

It is now generally accepted that the Papanicolaou and Traut monograph of 1943 was the catalyst that provoked a rediscovery of exfoliative cytology, a tool that had lain in disuse for 50 years.[16] As has been cited by Foot, the late 1940s and early 1950s marked a period of development for pulmonary cytology.[20] Numerous papers were written to report results of new techniques, ability to detect neoplastic cells, and prediction of histologic type of neoplasm. Of particular significance in these areas are the contributions of Wandall,[10] Bamforth,[21, 22] Woolner and McDonald,[23-30] Papanicolaou,[31-33] Farber and his associates,[12, 34-38] Clerf and Herbutt,[39-45] Foot,[20, 46] Umiker,[47-51] Richardson and his associates,[52, 53] and Koss.[54, 55]

REFERENCES

1. Papanicolaou, G. N.: New cancer diagnosis. *Proc Third Race Betterment Con*, p. 528, 1928.
2. Babes, A.: Diagnostic du cancer du col uterin par les frottis. *La Presse Medicale, 36:* 451–454, 1928.
3. Fontana, F.: Richerche fisiche sopra il veleno della vipera. J Guisti, Lucca, 1767.
4. Hughes, A.: *A History of Cytology*. Abelard-Schuman, London and New York, 1959.
5. Schleiden, M. J.: *Principles of Scientific Botany*. Trans. E Lankaster, London, 1849.

6. Schwann, T. and Schleiden, M. J.: *Mikroskopische Untersuchungen über die Uebereinstimmung in der struktur und dem Wachstum der Thiere und Pflanzen Muller's Archiv*, p. 137, 1838.
7. Donné, A.: *Cours de Microscopie Complementaire des Etudes Medicales Anatomie Microscopique et Physiologie des Fluides de L'Economie.* J. B. Bailliere, Paris, 1845.
8. Grunze, H.: A critical review and evaluation of cytodiagnosis in chest diseases. *Acta Cytol, 4:* 175–198, 1960.
9. Mueller, J.: *Ueber den feinern Bau und die Formen der Krankhaften Geschwulste.* G. Reimer, Berlin, 1838.
10. Wandal, H. H.: A study on neoplastic cells in sputum as a contribution to the diagnosis of primary lung cancer. *Acta Chir Scand, 91*(Suppl 93)*:* 1–143, 1944 (cit. by Wandal).
11. Walshe, W. H.: *Diseases of the Lungs.* London, 1843.
12. Farber, S. M., Benioff, M. A., Frost, J. K., Rosenthal, M., and Tibias, G.: Cytologic studies of sputum and bronchial secretions in primary carcinoma of the lung. *Dis Chest, 14:* 633–664, 1948.
13. Beale, L. S.: Examination of sputum from a case of cancer of the pharynx and adjacent parts. *Arch Med, 22:* 44–46, 1860.
14. Beale, L. S.: *The Microscope in Medicine.* J & A Churchill, London, 1879.
15. Russell, W. O., Neidhardt, H. W., Mountain, C. F., Griffith, K. M., and Chang, J. P.: Cytodiagnosis of lung cancer. A report of a four-year laboratory, clinical, and statistical study with a review of the literature on lung cancer and pulmonary cytology. *Acta Cytol, 7:* 1–44, 1963.
16. Papanicolaou, G. N. and Traut, H. F.: *The Diagnosis of Uterine Cancer by the Vaginal Smear.* Commonwealth Fund, New York, 1943.
17. Reagan, J. W.: Cellular pathology and uterine cancer. *Am J Clin Pathol, 62:* 150–164, 1974.
18. Hajdu, S. I.: Cytology from antiquity to Papanicolaou. *Acta Cytol, 21:* 668–676, 1977.
19. Dudgeon, L. S. and Wrigley, C. H.: On the demonstration of particles of malignant growth in the sputum by means of the wet-film method. *J Laryng Otol, 50:* 752–763, 1935.
20. Foot, N. C.: The identification of types of pulmonary cancer in cytologic smears. *Am J Pathol, 28:* 963–977, 1952.
21. Bamforth, J.: The examination of the sputum and pleural fluid in the diagnosis of malignant diseases of the lung. *Thorax, 1:* 118–127, 1946.
22. Bamforth, J. and Osborn, G. R.: Diagnosis from cells. *J Clin Pathol, 11:* 473–482, 1958.
23. Woolner, L. B. and McDonald, J. R.: Bronchogenic carcinoma: Diagnosis by microscopic examination of sputum and bronchial secretions; preliminary report. *Proc Staff Meetings Mayo Clin, 22:* 369–381, 1947.
24. Woolner, L. B. and McDonald, J. R.: Diagnosis of carcinoma of the lung; the value of cytologic study of sputum and bronchial secretions. *JAMA, 139:* 497–502, 1949.
25. Woolner, L. B. and McDonald, J. R.: Cytologic diagnosis of bronchogenic carcinoma. *Am J Clin Pathol, 19:* 765–769, 1949.
26. Woolner, L. B. and McDonald, J. R.: Carcinoma cells in sputum and bronchial secretions. *Surg Gynecol Obstet, 88:* 273–290, 1949.
27. Woolner, L. B. and McDonald, J. R.: Cytologic diagnosis of bronchogenic carcinoma. *Dis Chest, 17:* 1–10, 1950.
28. Woolner, L. B. and McDonald, J. R.: Cytology of sputum and bronchial secretions: studies on 588 patients with miscellaneous pulmonary lesions. *Ann Intern Med, 33:* 1164–1174, 1950.
29. McDonald, J. R.: Exfoliative cytology in genitourinary and pulmonary diseases. *Am J Clin Pathol, 24:* 684–687, 1954.
30. McDonald, J. R.: Pulmonary cytology. *Am J Surg, 89:* 462–464, 1955.
31. Watson, W. L., Cromwell, R., Craver, L., and Papanicolaou, G. N.: Cytology of bronchial secretions: Its role in the diagnosis of cancer. *J Thorac Surg, 18:* 113–122, 1949.
32. Papanicolaou, G. N. and Cromwell, H. A.: Diagnosis of cancer of the lung by the cytologic method. *Dis Chest, 15:* 412–418, 1949.
33. Papanicolaou, G. N. and Koprowska, I.: Carcinoma in situ of the right lower bronchus. *Cancer, 4:* 141–146, 1951.
34. Farber, S. M., McGrath, Jr., A. K., Benioff, M. A., and Rosenthal, M.: Evaluation of cytologic diagnosis of lung cancer. *JAMA, 144:* 1–4, 1950.
35. Farber, S. M., Rosenthal, M., Alston, E. F., Benioff, M. A., and McGrath, Jr., A. K.: *Cytologic Diagnosis of Lung Cancer.* Charles C Thomas, Springfield, Ill., 1950.
36. Farber, S. M. and Pharr, S. L.: The practicing physician and pulmonary cytology. *Lancet, 77:* 111–113, 1957.
37. Farber, S. M., Wood, D. A., Pharr, S. L., and Pierson, B.: Significant cytologic findings in non-malignant pulmonary disease. *Dis Chest, 31:* 1–13, 1957.
38. Farber, S. M.: Clinical appraisal of pulmonary cytology. *JAMA, 175:* 345–348, 1961.
39. Clerf, L. H. and Herbutt, P. A.: Diagnosis of bronchogenic carcinoma by examination of bronchial secretions. *Ann Otol, 55:* 646, 1946.

40. Herbutt, P. A. and Clerf, L. H.: Cytology of bronchial secretions: A diagnostic aid in the diagnosis of pulmonary tuberculosis. *Am Rev Tuberc, 54:* 488–494, 1946.
41. Herbutt, P. A. and Clerf, L. H.: Bronchogenic carcinoma. Diagnosis by cytologic study of bronchoscopically removed secretions. *JAMA, 130:* 1006–1012, 1946.
42. Herbutt, P. A.: Cancer cells in bronchial secretions. *Am J Pathol, 23:* 867–868, 1947.
43. Clerf, L. H. and Herbutt, P. A.: The value of cytological diagnosis of pulmonary malignancy. *Am Rev Tbc, 61:* 60–65, 1950.
44. Clerf, L. H. and Herbutt, P. A.: Early diagnosis of cancer of the lung. *JAMA, 150:* 793–795, 1952.
45. Herbutt, P. A.: Correlation of cytological with pathological findings in tumors of the lung. In *Proceedings of the Symposium on Exfoliative Cytology 1951.* American Cancer Society, New York, p. 50.
46. Foot, N. C.: Cytologic diagnosis in suspected pulmonary cancer. Critical analysis of smears from 1,000 persons. *Am J Clin Pathol, 25:* 223–240, 1955.
47. Umiker, W. O.: Cytology in bronchiogenic carcinoma. *Am J Clin Pathol, 22:* 558–563, 1952.
48. Umiker, W. O.: False-negative reports in the cytologic diagnosis of cancer of the lung. *Am J Clin Pathol, 28:* 37–45, 1957.
49. Umiker, W. O., DeWeese, M. S., and Lawrence, G. H.: Diagnosis of lung cancer by bronchoscopic biopsy, scalene lymph node biopsy, and cytologic smears. A report of 42 histologically proved cases. *Surgery, 41:* 705–713, 1957.
50. Umiker, W. O.: Diagnosis of bronchogenic carcinoma; an evaluation of pulmonary cytology, bronchoscopy and scalene lymph node biopsy. *Dis Chest, 37:* 82–90, 1960.
51. Umiker, W. O.: The current role of exfoliative cytopathology in the routine diagnosis of bronchogenic carcinoma; a five-year study of 152 consecutive, unselected cases. *Dis Chest, 40:* 154–159, 1961.
52. Richardson, H. L., Hunter, W. C., Conklin, W. S., and Petersen, A. B.: A cyto-histologic study of bronchial secretions. *Am J Clin Pathol, 19:* 323–327, 1949.
53. Richardson, H. L., Koss, L. G., and Simon, T. R.: An evaluation of the concomitant use of cytological and histocytological techniques in the recognition of cancer in exfoliated material from various sources. *Cancer, 8:* 948–950, 1955.
54. Koss, L. G. and Richardson, H. L.: Some pitfalls of cytological diagnosis of lung cancer. *Cancer, 8:* 937–947, 1955.
55. Koss, L. G.: Cellular changes simulating bronchogenic carcinoma. *Acta Unio Internat Contra Cancrum, 14:* 501–503, 1958.

CHAPTER
II
Cytopreparatory
Methods

As is true for most of the biomedical sciences which utilize morphology, excellent cytopreparatory techniques are mandatory for diagnostic accuracy. A specimen from the respiratory tract which has been prepared for cytologic diagnosis (1) should exhibit an abundance of well-preserved and stained diagnostic cellular material; (2) should have been prepared rapidly and with relative ease; (3) should remain preserved for permanent slide storage.

Many laboratories have studied techniques for the best realization of these criteria. The sectioning of blocks from sputum that has been embedded in paraffin is a technique known since the early part of this century.[1-3] Its use is still widespread, particularly in Europe. Undoubtedly pathologists who are not well grounded in the current principles of diagnostic cytopathology derive some comfort in the similarity of sections of sputum to tissue sections. Such comfort is built upon shaky foundations indeed, since diagnostic pickup of neoplastic cells is low and cellular morphology is blurred and distorted. Paraffin embedding and sectioning of sputum is, in the experience of the authors, the worst possible technical approach to respiratory cytology.

Various techniques for freeing and concentrating tumor cells by mucolysis were commendable in concept but were frequently too laborious and time consuming to be practical.[4-17]

Three major techniques have stood the test of time and are the most widely utilized today. They are the Saccomanno method,[18, 19] wet-film preparation, and membrane filtration.[20-25] These methods are used for spontaneously produced sputum, induced sputum,[26-42] and bronchial washings and brushings. The Papanicolaou method is the most generally accepted staining method.[43] In the United States in particular, it has gained virtually universal popularity.

FRESH MATERIAL FROM THE LOWER RESPIRATORY TRACT

For the first technique cited, an early-morning fresh, "deep cough" specimen of sputum is collected into a shallow, wide-mouthed sputum jar and brought immediately to the laboratory. It is examined grossly for tissue fragments, areas of fresh or old blood, or other fragments that could possibly be harboring tumor cells. Smears from these areas and others randomly sampled are prepared by gentle, even spreading of the specimen between two glass slides until a thin uniform smear is obtained. These are fixed immediately, without air drying, in 95% ethyl alcohol for 15 to 30 minutes. At the operating table direct smears are prepared from the bronchoscope and from the brush. The unfixed fresh bronchial washing is brought to the laboratory for membrane filter preparation. The brush is agitated in a balanced salts solution, and from this solution membrane filters are prepared.

PREFIXED MATERIAL FROM THE LOWER RESPIRATORY TRACT

In those situations where it is not possible to transmit unfixed material to the laboratory, prefixed sputum may be obtained by instructing the patient to expectorate into a sputum jar half filled with 70% ethyl alcohol. The theoretical advantages of this method are significantly offset by varying degrees of cellular preservation and a rubbery consistency of the sputum, which increases the difficulty of smear preparation.

THE SACCOMANNO METHOD FOR MATERIAL FROM THE LOWER RESPIRATORY TRACT

A technique for collection and prefixation of sputum, first reported by Saccomanno in 1963, is now gaining in popularity.[18] The Saccomanno method is a prefixation and concentration technique that can be used for both sputum and bronchial washings. Details of this method are provided in the appendix at the end of this chapter. The fixative already contained in the sputum bottle is either 50% ethyl or 50% isopropyl alcohol mixed with 2% polyethylene glycol, the liquid form of carbowax. This provides some preliminary fixation while the patient collects several samples of sputum either during several hours or even days. Because of the prefixation the specimen may be transported conveniently, making this method ideal for outpatient screening. Spontaneous or induced sputum may be submitted in this manner.

The cytology laboratory at the Mayo Clinic has recently modified the actual preparation of smears from prefixed and blended sputum as prepared by the Saccomanno technique. A vortex mixer is used to resuspend the blended and centrifuged specimen. A disposable glass pipette and rubber bulb may then be used to obtain several drops of resuspended cells and debris for preparation of the smears on glass slides. This makes it much easier to prepare uniform, evenly dispersed smears without thick clumps. Staining is thus more uniform, the cells appear to be more transparent, and screening is made easier. It should be remembered that the smears must be perfectly dry before they are placed in 95% ethyl alcohol for final fixation. The cells are protected by both the prefixation and the carbowax coating, so that drying artefacts do not occur.

With the amount of carbowax present in the Saccomanno fixative it has not been found necessary to use additional alcohol baths prior to staining. Staining clarity and sharpness are usually quite good with standard Papanicolaou techniques.[44]

Although the kitchen-type food blender may be used, there is some loss of material and some unevenness of blending with the large container supplied. The authors recommend the small-volume blender cup designed for handling specimens of limited quantity (see appendix). The result will be a more even preparation without thick clumps, which interfere with the preparation of smooth, uniform smears.

The blending method has been criticized because of its theoretical potential to destroy cells as a result of cutting by the blender blades. Documentation of this is completely lacking. Cell-seeding experiments fail to demonstrate any significant cell loss. Cells are not fragmented or destroyed by the blender, but convection currents are set up which produce a homogeneous mixture of cells and debris. Centrifugation also concentrates the cell material, with the resultant potential for a higher yield of abnormal cells. The technique has allowed Saccomanno to collect a large repository of cases demonstrating the developmental sequence, cytologically, of lung cancer. His results have confirmed those of the ongoing screening study of the high-risk male population of heavy cigarette smokers for lung cancer.[45]

A recent paper and an avalanche of letters have questioned the safety of this method in terms of exposure of the technologist to infectious aerosols during the blending step.[46-51] Although the authors do not doubt the theoretical objections expressed, experience with the method for more than 14 years has not produced any respiratory infections either in the authors' laboratories or that of Saccomanno. The exposure in the Medical College of Virginia laboratory has been 2,500 sputum specimens per technologist over the 14-year period. At least one-third of these specimens have come from patients with significant pulmonary infections, principally tuberculosis. Obviously, common sense should be observed in the handling of potentially infectious specimens. It is recommended that gloves, gown, and mask be worn during the preparation of sputum by this method. There is only theoretical and no practical documentation that a bacterial or fume hood is necessary.

Where clinical findings direct attention to a respiratory infection rather than a neoplasm, fresh sputum for cytologic examination is preferred. The blending technique may disperse and destroy the fragile hyphae of some of the mycotic organisms, but it seems to have no effect upon those organisms that grow as a yeast in tissue or on the cytopathologic effects produced by viruses. Bronchial brushing usually provides the most definitive material for diagnosis in those cases where there might be some objection to the use of the blending technique.

BRONCHIAL BRUSHING

Bronchial brushing has become an increasingly popular technique. It is performed most frequently by use of the flexible fiberoptic bronchoscope, allowing the operator to view the primary and lobar bronchi as well as many of the segmental bronchi.[52, 53] More important, the operator is able to visualize what is being brushed. The alternative technique is that of selective bronchial catheterization described by Fennessy.[54] Fluoroscopic control is necessary for this method and is also useful with the flexible fiberoptic bronchoscope.

Methods of smear preparation and fixation of bronchial brushing specimens are provided in the appendix. It is extremely important that these methods be followed if high-quality cytologic smears are to be obtained. The authors have found this is possible only if the cytotechnologist is present during the procedure to take the brush from the operator and prepare the smears. Points worthy of emphasis are the use of the rolling motion during smear preparation, wetting the slide with alcohol just prior to making the smear, and spreading the cells over only a small area (about the size of a dime) of the slide. This last technique reduces the amount of screening time. More importantly, it lays the cells out so that there is preservation of interrelationships, creating a tissue-like pattern. Consequently it is possible to render a more specific cytologic diagnosis in the case of neoplasms.

If it is not possible to have the cytotechnologist available for smear preparation, then it is better, after obtaining the sample, to cut the end from the disposable brush and place it in 5.0–10.0 ml of balanced salts solution for immediate transportation to the cytology laboratory. Smears prepared by the bronchoscopist or an untrained assistant are usually useless because of scattering and destruction of cells and the appearance of air-drying artefacts.

With the variety of methods available for handling and obtaining specimens from the respiratory tract for cytologic diagnosis, it should be remembered that the type of specimen may affect the cytologic findings from the same neoplasm. Hence poorly differentiated squamous-cell carcinoma of the lung has a different cytomorphology in bronchial washings, brushing, and sputum. The authors will attempt to point out these differences throughout this monograph, as they may be a source of confusion in the reporting of cytologic results in histopathologic terminology.

REFERENCES

1. Mandlebaum, F. S.: The diagnosis of malignant tumors by paraffin sections of centrifuged exudates. *J Lab Clin Med, 2:* 580, 1971.
2. Wihman, G. and Bergstrom, J.: Histological technique for the examination of the cell content of sputum. *Acta Med Scand, 142:* 433–440, 1952.
3. Abramson, W., Dzenis, V., and Hicks, S.: Cytologic study of sputa and exudates using paraffin tubes. *Acta Cytol, 8:* 306–310, 1964.
4. Haynes, E.: Trypsin as a digestant of sputum and other both fluids preliminary to examination for acid-fast bacilli. *J Lab Clin Med, 27:* 806–809, 1942.
5. Farber, S. M., Pharr, S. D., Wood, D. A., and Gorman, R. D.: The mucolytic and digestive action of trypsin in the preparation of sputum for cytologic study. *Science, 117:* 687–690, 1953.
6. Rubin, C. E. and Benditt, E. P.: A simplified technique using chymotrypsin lavage for the cytological diagnosis of gastric cancer. *Cancer, 8:* 1137–1141, 1955.
7. Pharr, S. L., Farber, S. M., and King, E. B.: Cellular concentration of sputum for cytologic examination. In *Trans Fifth Annual Meeting of the Inter Society Cytology Council*, p. 65, 1957.
8. Umiker, W., Young, L., and Waite, B.: The use of chymotrypsin for the concentration of sputum in the cytologic diagnosis of lung cancer. *Univ Mich Med Bulletin, 24:* 265, 1958.
9. Rastgeldi, S., Tomenius, J. A., and Williams, G.: The simultaneous separation and concentration of corpuscular elements and bacteria from sputum. *Acta Cytol, 3:* 183–187, 1959.
10. Umiker, W. and Sourenne, R.: A simple method for concentrating carcinoma cells in sputum. *Amer J Clin Pathol, 35:* 411–412, 1961.
11. Pharr, S. M. and Farber, S. M.: Cellular concentration of sputum and bronchial aspirations by tryptic digestion. *Acta Cytol, 6:* 447–454, 1962.
12. Takahashi, M. and Urabe, M.: A new cell concentration method for cancer cytology of sputum. *Cancer, 16:* 199–204, 1963.
13. Knudtson, K. P.: Mucolytic action of hyaluronidase on sputum for the cytological diagnosis of lung cancer. *Acta Cytol, 7:* 59–61, 1963.
14. Liu, W.: Concentration and fractionation of cytologic elements in sputum. *Acta Cytol, 10:* 368–372, 1966.
15. Takahashi, M., Hashimoto, K., and Osada, H.: Parenteral administration of chymotrypsin for the early detection of cancer cells in sputum. *Acta Cytol, 11:* 61–63, 1967.
16. McCarty, S. A.: Solving the cytopreparation problem of mucoid specimens with a mucoliquifying agent (Mucolexx) and nucleopore filters. *Acta Cytol, 16:* 221–223, 1972.
17. Bonime, R. G.: Improved procedure for the preparation of pulmonary cytology smears. *Acta Cytol, 16:* 543–545, 1972.

18. Saccomanno, G., Saunders, R. P., Ellis, H., Archer, V. E., Wood, B. G., and Beckler, P. A.: Concentration of carcinoma or atypical cells in sputum. *Acta Cytol, 7:* 305–310, 1963.
19. Ellis, H. D. and Kernosky, J. J.: Efficiency of concentrating malignant cells in sputum. *Acta Cytol, 7:* 372–373, 1963.
20. Haley, L. D. and Arch, R.: Use of millipore membrane filter in the diagnostic tuberculosis laboratory. *Amer J Clin Pathol, 27:* 117–121, 1957.
21. Chang, J. P., Anken, M., and Russell, W. O.: Sputum cell concentration by membrane filtration for cancer diagnosis: A preliminary report. *Acta Cytol, 5:* 168–172, 1961.
22. Chang, J. P., Anken, M., and Russell, W. O.: Liquifaction and membrane filtration of sputum for the diagnosis of cancer. *Am J Clin Pathol, 37:* 584–592, 1962.
23. Chang, S. C. and Russell, W. O.: A simplified and rapid filtration technique for concentrating cancer cells in sputum. *Acta Cytol, 8:* 348–349, 1964.
24. Fields, M. J., Martin, W. F., Young, B. L., and Tweeddale, D. N.: Application of the Nedelkoff-Christopherson millipore method to sputum cytology. *Acta Cytol, 10:* 220–222, 1966.
25. Suprun, H.: A comparative filter technique study and the relative efficiency of these sieves as applied in sputum cytology for pulmonary cancer cytodiagnosis. *Acta Cytol, 18:* 248–251, 1974.
26. Bickerman, H. A., Sproul, E. E., and Barach, A. L.: An aerosol method of producing bronchial secretions in human subjects: A clinical technique for detection of lung cancer. *Dis Chest, 33:* 347–362, 1958.
27. Allan, W. B., Whittlesey, P., Haroutunian, L. M., and Kelly, E. B.: The use of sulfur dioxide as a diagnostic aid in pulmonary cancer. Preliminary report. *Cancer, 11:* 938, 1958.
28. Sproul, E. E.: Cytology of induced sputum as a diagnostic tool. In *Transactions of the Sixth Annual Meeting of the Intersociety Cytology Council.* New York, 1958, p. 145.
29. Sproul, E. E.: Superheated aerosol induced sputum in detection of lung cancer in hospital practice. In *Transactions of the Seventh Annual Meeting of the Intersociety Cytology Council.* Detroit, Michigan, 1959, p. 290.
30. Rome, D. S.: Value of aerosol-produced sputum as screening technique for lung cancer. *Acta Unio Internat Contra Cancrum, 15:* 474, 1959.
31. Barach, A. L., Bickerman, H. A., Beck, G. L., Nanda, K. G. S., and Pons, E. R.: Induced sputum as a diagnostic technique for cancer of the lungs. *Arch Intern Med, 106:* 230–236, 1960.
32. Berkson, D. M., and Snider, G. L.: Heated hypertonic aerosol in collecting sputum specimens for cytological diagnosis. *AMA, 173:* 135–138, 1960.
33. Umiker, W. O., Korst, D. R., Cole, R. P., and Manikas, S. G.: Collection of sputum for cytologic examination. Spontaneous vs. artificially produced sputum. *N Engl J Med, 262:* 565–566, 1960.
34. Umiker, W. O.: A new vista in pulmonary cytology; aerosol induction of sputum. *Dis Chest, 39:* 512–515, 1961.
35. Olson, R. G., Froeb, H. F., and Palmer, L. A.: Sputum cytology after inhalation of heated propylene glycol. *JAMA, 178:* 668–670, 1961.
36. Rome, D. S.: Value of aerosol-produced sputum as screening technic for lung cancer. *N Y State J Med, 61:* 2054–2060, 1961.
37. Leilop, L., Garret, M., and Lyons, H. A.: Evaluation of technique and results for obtaining sputum for lung carcinoma screening. A study by blind technique. *Am Rev Respir Dis, 83:* 803–807, 1961.
38. Rome, D. S. and Olson, K. B.: A direct comparison of natural and aerosol produced sputum collected from 776 asymptomatic men. *Acta Cytol, 5:* 173–176, 1961.
39. Brenner, S. A., Lambert, R. L., and Pablo, G. E.: Superheated aerosol induced sputum in the cytodiagnosis of lung cancer. *Acta Cytol, 6:* 405–408, 1962.
40. Sproul, E. E., Huvos, A., and Britsch, C.: A two-year follow-up study of 261 patients examined by use of superheated aerosol induced sputum. *Acta Cytol, 6:* 409–412, 1962.
41. Roberts, T. W., Pollak, A., Howard, R., and Howard, E.: Tracheo-bronchial cytology utilizing an improved tussilator (cough machine). *Acta Cytol, 7:* 174–179, 1963.
42. Tweeddale, D. N., Harbord, R. P., Nuzum, C. T., Pielemeier, B., and Kington, E.: A new technique to obtain sputum for cytologic study: External percussion and vibration of the chest wall. *Acta Cytol, 10:* 214–219, 1966.
43. Papanicolaou, G. N.: A new procedure for staining vaginal smears. *Science, 95:* 2469: 438–439, 1942.
44. Keebler, C. M., Reagan, J. W., and Wied, G. L. (Eds): *Compendium on Cytopreparatory Techniques.* Tutorial proceedings, third ed., Tutorials of Cytology, Chicago, 1974.
45. Saccomanno, G.: Personal communication.
46. O'Hara, C. M. and Birmingham, S. P.: Sputum fixative: How safe is 50 percent alcohol? *Acta Cytol, 20:* 400–403, 1976.
47. Elefson, D. E.: Letter to the Editor. *Acta Cytol, 21:* 494–495, 1977.
48. Harris, M. J.: Letter to the Editor. *Acta Cytol, 21:* 493, 1977.
49. Mitchell, P. D.: Letter to the Editor. *Acta Cytol, 21:* 493, 1977.
50. Saccomanno, G.: Letter to the Editor. *Acta Cytol, 21:* 495, 1977.

51. Tucker, B.: Letter to the Editor. *Acta Cytol, 21:* 492–493, 1977.
52. Marsh, B. R., Frost, J. K., Erozan, Y. S., Carter, D., and Proctor, D. F.: Flexible fiberoptic bronchoscopy. Its place in the search for lung cancer. *Ann Otol Rhinol Laryngol, 82:* 757–764, 1973.
53. Zavala, D. C.: Diagnostic fiberoptic bronchoscopy: Technique and results of biopsy in 600 patients. *Chest, 68:* 12–19, 1975.
54. Manalo-Estrella, P., Fry, W. A., and Davidson, D.: Selective bronchial catheterization and brushing. *Manual of Dept. of Pathology and Surgery*, Evanston Hospital and Northwestern University, Evanston, Illinois.

APPENDIX

I. COLLECTION TECHNIQUES

A. Sputum without fixation

1. An early-morning fresh "deep cough" specimen is collected into a wide-mouthed sputum jar.
2. This fresh, unfixed specimen is brought immediately to the laboratory.
3. On gross examination areas thought to contain tissue fragments, areas of fresh or old blood, or other areas thought to harbor tumor cells are chosen for smears.
4. Smears are prepared by gentle, even spreading of selected portions of the specimen between two glass slides until a thin, even smear is obtained.
5. Smears are fixed immediately in 95% ethyl alcohol. No air drying is permissible.
6. After a minimum of 15 min fixation in 95% alcohol, smears are ready for staining by Papanicolaou technique.
7. Bronchial washings may be prepared by the same technique, but in addition two filters (Millipore or Nuclepore) from diluted bronchial washings are made. Filters are made first after dilution of the bronchial washings with balanced salts solution. The remainder of the specimen is then centrifuged and direct smears are made from the cell button. Totally frosted slides are preferable for these smears.

EQUIPMENT AND SUPPLIES

1. Balanced Salts Solution, Polyonic^R R 148, Cutter Laboratories Inc., Berkeley, California 94710 or Plasma-Lyte 148 in water, Travonal Laboratories, Inc., Morton Grove, Illinois 60053.
2. Dakin All-Frosted Microscopic Slides, Cat. No. V58955, Aloe Scientific, St. Louis, Missouri 63103.

B. Sputum with fixation (Saccomanno technique)

1. Collect first-morning specimens from deep cough over three successive days, or combine several specimens collected over a 6–8-hr period in the same container.
2. Collecting container should be wide-mouth glass or plastic vessel containing 50.0 ml of 50% ethyl or isopropyl alcohol and 2% polyethylene glycol (carbowax).
3. Entire specimen is placed in a small-volume blending cup adapted to a household-type food blender.
4. Blend at the highest speed for 10 sec. Be sure the blending-vessel cap is securely in place.
5. In 50.0 ml centrifuge tubes, centrifuge the blended specimen for 10 min at minimum of 1,000 rpm.

6. Handling each centrifuge tube one at a time, pour off the supernatant, leaving a small volume (not more than 1.0 ml) in the tube with the cell button and debris.
7. Use a vortex mixer to blend this cell material and residual fixative.
8. With a disposable pipette place a drop of blended cell suspension on a clear glass slide with a frosted end. Be sure slides are labeled for patient identification.
9. Use a similar slide inverted against the first one, and, as the drop spreads between the two slides, pull them apart to make two even smears.
10. It may occasionally be necessary to work the slides back and forth to get even smears. Do not apply too much pressure to the slides as this will cause disruption and distortion of the cells.
11. Allow prepared smears to dry completely before placing in 95% ethyl alcohol for final fixation.
12. After the minimum of 15 min final fixation in 95% alcohol, slides are ready for staining by the Papanicolaou technique.
13. The same blending procedure may be used for bronchial washings, if collected in the 50% alcohol and carbowax fixative. It is not recommended for bronchial washings submitted unfixed in balanced salts solution.
14. For prefixed or unfixed bronchial washings, the authors prefer the use of totally frosted slides in the preparation of direct smears.
15. Prior to blending of bronchial washings or preparation of direct smears, make two filters (Millipore or Nuclepore) from diluted bronchial washings. Dilute bronchial washings with balanced salts solution. Dilute until nearly clear in appearance.

EQUIPMENT AND SUPPLIES

1. Carbowax 1540 (polyethylene glycol), Cat. No. 08047. Applied Science Laboratories, State College, Pennsylvania, 16801, or Union Carbide Corp. Chemical and Plastic Sales, Beltway Building, 6707 Whitestone Rd., Baltimore, Maryland, 21207.
2. 50.0 ml plastic tubes, conical graduated with screw cap. Falcon Plastics (Division Bio-Quest), 1950 Williams Drive, Oxnard, California, 93030. Other distributors: Fisher Scientific or Curtin Scientific.
3. 250.0 ml Eberback semimicro containers, Scientific Products, 8055 McGaw Rd., Columbia, Maryland, 21045.
4. Waring two-speed blender (or other type food blender), Waring Products Corp., Winsted, Connecticut, Model RL-6. Distributed by Scientific Products.
5. Electric mixer (Genie Vortex). Distributed by Scientific Products.
6. Clear polystyrene jars, 8 oz capacity, 89 mm thread, 3⅛ × 2¼ in. Parkway Plastics, Inc., 561 Stelton Rd., Piscataway, New Jersey, 08854.
7. Saliva test-jar mailer, die-cut mailer, length 3⅞ in., width 3⅞ in., depth 2⅞ in. Mullen test, 200B Snobrite, printed. Packing Corp. of America. 1100 Whistler Ave., Baltimore, Maryland, 21233.

C. Bronchial brushings

1. Technique of selective bronchial catheterization—Fennessy
   a. Premedication is provided by parenteral morphine and pentobarbital but should be at the discretion of the bronchoscopist.
   b. Local anesthesia is obtained in the mouth and pharynx with 2% tetracaine and in the larynx, trachea and bronchi with 5% cocaine.
   c. A small rubber catheter is passed through the mouth and into the trachea. The patient is positioned under the fluoroscope.
   d. A Seldinger guide wire is advanced through the cathether into the trachea and the initial catheter is removed. A premolded plastic catheter is then passed over the guide wire and selectively maneuvered into the area for brushing with fluoroscopic control.
   e. The guide wire is removed and a disposable Fennessy bronchial brush is advanced into position, and a specimen is obtained.
   f. Before the brush is withdrawn, premoisten several Dakin (totally frosted) slides with 95% ethyl or isopropyl alcohol. Shake off the excess alcohol.

    g. Withdraw the brush and make several smears on the moistened slides. Use a *gentle* rolling motion with the brush over an area in the center of the slide about the size of a dime. Fix immediately while still wet in 95% ethyl or isopropyl alcohol.

    h. Speed is important in preparing the smears, as drying artefacts occur rapidly.

    i. After the smears are made, shake the brush in 10.0 ml of balanced salts solution in a test tube or culture tube.

    j. Make several filters (Millipore or Nuclepore) from the cells dislodged into the balanced salts solution.

    k. Several brushings may be taken from the same or different areas. The disposable brush tips may be cut off and sent for culture in suitable transport media.

    l. Tissue fragments seen on the brush tips may be processed as a biopsy after fixation in formalin or Zenker's fluid.

    m. Selective catheter irrigations are usually performed with sterile balanced salts solution immediately after brushing.

    n. A selective bronchogram may then be performed if indicated.

EQUIPMENT AND SUPPLIES

1. Plastic bronchial catheters are made from KIFA catheter tubing (size 8F) and obtained from U.S. Catheter and Instrument Corp., Glens Falls, New York. Fennessy bronchial brushes are obtained from the Mill-Rose Company, Mentor, Ohio.
2. Disposable brushes and catheter may also be obtained from Electro-Catheter Corp., Rahway, New Jersey, 07065. Order Elecath, Cat. No. BBS, french size 8, length 95 cm.

2. Technique using the fiberoptic bronchoscope
    a. Premedication and topical anesthesia is the same as that used in the technique for selective catheterization.

    b. The fiberoptic bronchoscope may be passed through the nose, mouth, or tracheostomy stoma.

    c. Smears and specimens for culture are obtained and prepared in the same manner as described under selective bronchial catheterization.

    d. It is equally advantageous with this technique to have a cytotechnologist available at the time of brushing to make the smears and prepare the specimen in balanced salts solution for filter preparation.

    e. With both techniques, notes of lesions seen and the history of the case should be obtained.

EQUIPMENT AND SUPPLIES

1. Olympus BF-5B bronchofiberscope, Olympus Corp. of America, New Hyde Park, New York.

II. LABORATORY PREPARATION PROCEDURES

A. Millipore filtration

1. Identify the filter by writing the last three digits of the specimen accession number twice on the margin of the filter.
2. Expand the filter by submerging it in a petri dish of 95% ethyl alcohol for about 1 min.
3. Place the filter on the filter support base and then secure by placing funnel on top.
4. Add a balanced salts solution to a depth of approximately 4–5 cm to the funnel. Filter through a small amount of this solution to clear the filter of 95% alcohol.
5. Turn on the vacuum and slowly add the specimen, then repeatedly rinse the filter by adding balanced salts solution. Maintain a reservoir depth of at least 1 cm.
6. Add 95% ethyl alcohol to reservoir and filter through. When 95% ethyl alcohol level is about 1 cm turn off the vacuum and let cells fix for about 1 min. Turn on the vacuum

again and filter until just a small amount of 95% ethyl alcohol remains over the filter. Do not permit filtration to proceed to drying of the filter.
7. Remove the filter and place in a petri dish of 95% ethyl alcohol. Allow to fix in the petri dish for an additional ½ hr before staining.

EQUIPMENT AND SUPPLIES
1. Millipore filtering equipment, Millipore Corporation, Bedford, Massachusetts, 01730. Millipore cat. no. includes stainless steel filter holder, funnel, base, filter support screen, stopper, and Teflon gasket. An all-glass filter holder is also available. Filters are 47 mm, 5.0 micron pore size. Use three-place filter holder manifold or flask with side arm. Millipore forceps. Stainless steel clips.

B.   Staining of Millipore filters
1. Filters may be taken through staining in one of two methods.
    a. Using bulldog clips and applicator sticks.
        i. With millipore forceps grasp filter on margin away from numbers.
        ii. Attach bulldog clip to numbered margin of filter and insert applicator sticks (2)` through hole in clip.
        iii. Suspend filter in 95% ethyl alcohol so all of filtered area is in solution.
    b. Using polyethylene holder* and stainless steel clips.
        i. Attach filter to holder with stainless steel clips at the edges at nine, three, and six o'clock.
        ii. Place holder in staining racks and carry through staining setup.

C. Modified Papanicolaou staining procedure for Millipore filters

1. Tap water	10 dips
2. Tap water	10 dips
3. Hematoxylin—Gill	1 min
4. Tap water	10 dips
5. Tap water	10 dips
6. 0.05% hydrochloric acid	30 sec to 1 min until filter is colored light yellow
7. Tap water	10 dips
8. Tap water	10 dips
9. Scott's tap water	1 min
10. Tap water	10 dips
11. Tap water	10 dips
12. 95% ethyl alcohol	10 dips
13. 95% ethyl alcohol	10 dips
14. Orange G	1 min
15. 95% ethyl alcohol	1 min, but do not dip filters; let them sit
16. 95% ethyl alcohol	1 min sitting
17. 95% ethyl alcohol	1 min sitting
18. Eosin polychrome	8 min
19. 95% ethyl alcohol	4 min sitting
20. 95% ethyl alcohol	2 min sitting
21. 95% ethyl alcohol	1 min sitting
22. Isopropyl alcohol	1 min sitting
23. Isopropyl alcohol	1 min sitting
24. Xylene	1 min
25. Xylene	1 min
26. Xylene	Until coverslipped

* Holders are cut from thin polyethylene plastic suitable to fit in ordinary slide racks. A central hole is cut slightly smaller than the size of the filter to be stained.

D. Modified Papanicolaou staining procedure for smears of sputum and from bronchial brushes

1.	Tap water	10 dips
2.	Tap water	10 dips
3.	Hematoxylin—Gill	2 min
4.	Tap water	10 dips
5.	Tap water	10 dips
6.	Tap water	10 dips
7.	Tap water	10 dips
8.	Scott's tap water	1 min
9.	Tap water	10 dips
10.	Tap water	10 dips
11.	95% ethyl alcohol	10 dips
12.	95% ethyl alcohol	10 dips
13.	Orange G	1 min
14.	95% ethyl alcohol	10 dips
15.	95% ethyl alcohol	10 dips
16.	95% ethyl alcohol	10 dips
17.	Eosin polychrome	6 min
18.	95% ethyl alcohol	10 dips
19.	95% ethyl alcohol	10 dips
20.	95% ethyl alcohol	10 dips
21.	100% ethyl alcohol	10 dips
22.	100% ethyl alcohol	10 dips
23.	Xylene	10 dips
24.	Xylene	10 dips
25.	Xylene	Until coverslipped

E. Mounting of cellular preparations

Mounting of smears may be accomplished with any of the standard commercially available mounting media. Use #1 glass coverslips for both smears and filters. Mounting of Millipore filters for optimum results requires mounting media of the same refractive index as the filter. Two mounting media are currently available from Eukitt and Kleermount. The procedure for mounting filters is as follows:
1. Pour mounting media (about 20.0 ml) into aluminum foil milk-testing cups.
2. Cut the stained filter in two between the numbers at the edge and trim the filter close to the stained area.
3. Place the filters cell side up in the milk cups. Let sit for 5 min.
4. With millipore forceps, pick up filter by numbered end. Let excess mounting media drain off.
5. Gently lay filter onto 3 × 1-in. clear glass slide.
6. Avoid trapping air bubbles underneath filter by rolling fiber onto the slide.
7. Place coverslip over filter without using any extra mounting media. Do not drain extra mounting media from edge of the slide.

EQUIPMENT AND SUPPLIES
1. Eukitt mounting media available from Calibrated Instruments Inc., 731 Saw Mill River Road., Ardsley, New York, 10502.
2. Kleermount mounting media available from Carolina Biological Supply Co., Burlington, North Carolina, 27215.

F. Preparation of stains

Gill's half-oxidized hematoxylin[1a] is used in the authors' laboratories for staining of smears and filters. With this hematoxylin, an acid decolorization step is necessary for filters but not for smears. This step removes the hematoxylin from the filter. Other types of hematoxylin also need an acid decolorization step to eliminate background color from the filter. *It is important to wash filters in clear water after hematoxylin staining* rather than just dipping them in several changes of water, so that they will not wash in a mixture of hematoxylin and water. Solutions, particularly alcohols and water, need to be changed frequently and filtered, depending upon the volume of material being processed. Quality control of the hematoxylin step can be achieved by removing one slide or filter after hematoxylin staining and washing and checking the degree of nuclear staining under the microscope. Appropriate corrections are more easily made at that time.

1. *Hematoxylin—Gill*

Distilled water	730.0 ml
Ethylene glycol	250.0 ml
Hematoxylin, anhydrous, C.I. #75290	2.0 g
Sodium iodate	0.2 g
Aluminum sulfate $AL_2(SO_4)_3 \cdot 18H_2O$	17.6 g
Glacial acetic acid	20.0 ml

These chemicals should be combined in the order listed, and should be stirred for 1 hr on a magnetic mixer at room temperature. The stain can be used immediately.

2. *Scott's tap water*

Magnesium sulfate, anhydrous	10.0 g
Sodium bicarbonate	2.0 g
Distilled or tap water	1.0 liter

3. *Orange G*

   Stock solution: Prepare a 0.2M orange G aqueous solution by dissolving 9.05 g orange G in distilled water, making up to 100.0 ml in a volumetric flask. Make sure to correct for dye content. Check the manufacturer's label for percentage dye content.

   Working solution: Combine the following ingredients in a flask at room temperature:

0.2M orange G stock solution, C.I. #16230	10.0 ml
Phosphotungstic acid	1.0 g
95% ethyl alcohol	995.0 ml
Glacial acetic acid	4.0 ml

   The solution may be used immediately and should be filtered before use.

4. *Eosin polychrome*

   Stock solution: Prepare individually the following aqueous stock solutions by dissolving the dyes in distilled water at 70–80°C, making up to 100.0 ml in separate volumetric flasks. Remember to correct for dye content.

0.05M light-green SF yellowish, C.I. #42095	3.96 g
0.30M eosin Y, C.I. #45380	20.8

   Working solution: Combine the following ingredients in a flask at room temperature.

0.05M light green SF yellowish	10.0 ml
0.30M eosin Y	20.0 ml
Phosphotungstic acid	2.0 g
95% ethyl alcohol	700.0 ml
Absolute methyl alcohol	250.0 ml
Glacial acetic acid	20.0 ml

   The stain may be used immediately and should be filtered before use.

   Gill hematoxylin is available from Lerner Laboratories, 207 Greenwich Ave., Stamford,

Connecticut, 06902. The other stains are available from a variety of commercial suppliers. Check the formulation before using in this staining schedule.

III. FINE-NEEDLE ASPIRATION BIOPSY—TRANSTHORACIC

The patient should be examined and the X-ray films and history reviewed. The authors prefer to perform this procedure in collaboration with a thoracic surgeon in an operating room. It can be performed at the bedside or on outpatients if absolutely necessary. The patient is usually positioned sitting up and a skin-marking pencil is used to determine the site for puncture. This should be selected so as to traverse as little soft tissue and lung as possible to reach the lesion. The skin in the area of puncture is cleaned with alcohol, and 1% or 2% xylocaine or lidocaine anesthesia is infiltrated into skin and soft tissue, carrying this local anesthesia down to the pleura. About 5.0–10.0 ml will usually suffice.

The skin area is then prepared with sterile solution of Betadine, and the aspirating needle is introduced through the skin in the predetermined plane of entry into the lung lesion. The needle should be attached before introduction to a 20.0 ml Luer loc syringe, or else a needle with a stylus should be used (see Equipment and Supplies).

The patient may experience some pain at the time the needle passes through the pleura and into the lung. If the lesion is not flush with the chest wall in the plane of entry, there is very little resistance to advancement of the needle through normal lung. Pulling back slightly on the syringe to see if air is withdrawn may confirm that the needle is in air-containing lung. The lesion is found by feeling differences in resistance and judging the depth of the lesion from the films and the amount of needle remaining above the skin surface. For small lesions fluoroscopic control is very useful. During fluoroscopy, if the lesion is entered, it will dance on the fluoroscope when the needle is moved very slightly.

Once the needle is felt to be in the lesion the aspirating pistol should be attached to the syringe. With full vacuum applied the needle is moved back and forth over a short distance within the lesion and may also be rotated. It is important to watch the point at which the syringe joins the needle for the appearance of any material and/or blood. When blood is visualized or after 10 to 12 passes at the lesion, the aspiration is stopped by allowing the vacuum in the syringe to equate. The needle is then slowly withdrawn and pressure applied to the puncture site with a sterile gauze pad.

Smears are prepared from a small drop of the aspirate placed on a glass slide. This is accomplished by detaching the needle from the syringe and filling the syringe with air. Reattach the needle and advance the plunger of the syringe to express a small drop of the aspirate on the center of a glass slide. Press the bevel of the needle against the slide while expressing the drop so that there is no intervening air space, which will cause drying artefacts in the alcohol-fixed smears.

Make the smear by using a coverglass or another slide laid on top of the drop and pulled apart quickly as the drop spreads from the weight of the coverglass or slide. The principle is the same as that in making bone marrow smears. The smears should occupy a small area of the slide; this gives a simulated tissue pattern as well as good cytologic detail. Fix some of the smears immediately in 95% ethyl alcohol for Papanicolaou staining. Air-dry the remaining smears for staining by other methods.

Transthoracic aspirates may be particularly bloody and 1.0 ml or somewhat more material may be obtained. Smears are prepared in the same manner but the remaining blood should be allowed to clot in a watch glass, fixed with suitable tissue fixative, and prepared as a cell block using Harris'[2a] method.

Staining Techniques for Aspiration Specimens

1. *Papanicolaou:* Any of the modifications available for staining smears are satisfactory.
2. *Metachrome B:* Azure A, French, or MacNeal formula. To four parts of 1% aqueous

solution azure A (azure 1), which has been previously filtered, add very rapidly one part of filtered 0.5% aqueous Erie Garnet B. The mixture is immediately filtered to prevent precipitation. Occasional refiltering is necessary if the mixture has been standing a month or more.

Staining Procedure: The air-dried smears are placed in a Coplin jar with the stock stain (which has been diluted with an equal volume of distilled water) for 8–10 sec. Wash gently a few seconds in running tap water. The smears may be examined wet or mounted with a coverslip using a few drops of 40% glucose solution (Brun's media). The smears may be made semipermanent by soaking off the coverslip in water and allowing them to dry completely. When dry, dip in xylene for several seconds and coverslip with any of the commercially available slide-mounting media. The stain fades after about five years.

3. *Diff-Quik^R:* This stain is available as a commercial kit and is a modified Wright stain. It is a three-solution, three-step method that is both fast and practical. It gives good cellular detail and is comparable to May-Grunwald-Giemsa and Wright-Giemsa.

Staining Procedure: The air-dried smears are dipped for five seconds (five dips) in solutions I, II, and III, in that order, draining the excess stain from the slides between solutions. After the third solution the slide is rinsed with water and allowed to dry or examined wet. After complete drying the slide may be made permanent by immersion in xylene for several seconds and mounting with any of the commercial tissue-mounting media and a coverslip. For more intense staining increase the number of dips, but in any case, use not less than three full, 1-sec dips. To increase eosinophilia increase the number of dips in solution I. To increase basophilia increase the number of dips in solution II. Solutions should be kept tightly covered when not in use.

4. *Modified May-Grunwald-Giemsa:* May-Grunwald stock stain is prepared by dissolving 1.0 g eosin-methylene blue in 100.0 ml of methyl alcohol. The working stain is prepared by adding 40.0 ml of stock stain to 20.0 ml of methyl alcohol in a Coplin jar. Giemsa stock stain is prepared by dissolving 1.0 g Giemsa powder in 66.0 ml of glycerin and incubating at 37°C for 3 hr. The working Giemsa stain is prepared by adding 5.0 ml of stock stain to 45.0 ml of distilled water in a Coplin jar.

Staining Procedure: Immerse the air-dried aspiration smears in working May-Grunwald stain for 15 min. Rinse gently in tap water. Immerse the smears in working Giemsa stain for 15 min. Rinse gently with tap water. Allow to air-dry. Dip in xylene for 10 sec and mount with any commercial tissue-mounting media. Prepare working May-Grunwald stain fresh once per week. Prepare working Giemsa stain fresh daily. The stock Giemsa stain is good for six months if refrigerated. The stock May-Grunwald stain is good indefinitely and does not require refrigeration.

5. *Modified Wright-Giemsa:* Any formula for Wright stain in standard laboratory use is satisfactory. Prepare stock Giemsa stain as previously described. Two buffer solutions are required, pH 6.4 and pH 6.8. These can be purchased in dry-pack form commercially. Prepare Giemsa working stain by diluting one part Giemsa stock stain with nine parts of buffer solution, pH 6.8. Prepare this fresh each day.

Staining Procedure: The air-dried aspiration smears are flooded with methyl alcohol and allowed to evaporate to dryness. Then flood the smears with Wright's stain for 3 min. Add drop by drop to the Wright's stain on the smears, buffer solution pH 6.4, blowing on the slide to mix the stain and buffer. The stain should develop a green sheen when adequate buffer has been added. Allow to stand for 4 min. Wash in tap water and dry completely. Stain with working Giemsa solution for 3 min. Wash with tap water and dry completely. Dip in xylene for 10 sec and mount in any commercial tissue-mounting media with a coverslip.

6. *Hematoxylin and Eosin:* The procedure is that of Hajdu and Melamed.[3a] Harris formula hematoxylin and eosin Y are used, both of which are commercially available.

EQUIPMENT AND SUPPLIES

1. Cameco Syringe Pistol[R], available from Precision Dynamics Corp., 3031 Thornton Ave., Burbank, California, 91504.
2. Aspir-Gun,[R] available from Everest Co., 5 Sherman St., Linden, New Jersey, 07036.
3. 22 gauge, 0.6 mm external diameter 6- and 7-in. needles available from Becton Dickinson Division of Becton Dickson & Co., Rutherford, New Jersey, 07070. Catalogue # SH-2686 6" and 7".
3. 22 gauge, 15.0 and 20.0 cm needle with stylus, available as CHIBA Biopsy CHN 22, 15.0 and 20.0 cm needle from C.I. Cook Inc., P. O. Box 489, Bloomington, Indiana, 47401.
4. Azure A, available from MCB Manufacturing Chemists, 2902 Highland Ave., Norwood, Ohio, 45202, Catalogue No. AX 1875.
5. Erie garnet B, available from City Chemical Co., 132 W. 22nd. St., New York, New York, 10011.
6. Diff-Quick,[R] available from Harleco, Division of American Hospital Supply Corp., 480 Democrat Rd., Gibbstown, New Jersey, 08027.
7. May-Grunwald stock stain available as Jenner stain from Paragon C & C Co., Inc., 190 Willow Ave., Bronx, New York, 10454.
8. Giemsa stain available as the tissue stain Wolbach modification or the blood stain, from Harleco. Either is satisfactory.
9. Wright's stain available in dry-pack form from Harleco.
10. Buffer solutions available as the salt in dry-pack form from Harleco.
11. Harris hematoxylin in solution available from Harleco.
12. Eosin Y aqueous 5% pure dye solution available from Harleco.

REFERENCES FOR APPENDIX

1a. Gill, G. W., Frost, J. K., and Miller, K. A.: A new formula for a half-oxidized hematoxylin solution that neither overstains nor requires differentiation. *Acta Cytol, 18:* 300–311, 1974.
2a. Harris, M. J. and Keebler, C. M.: Cytopreparatory Techniques. In *Compendium On Cytopreparatory Techniques*. 3rd ed. Keebler, C. M., Reagan, J. W., and Wied, G. L. eds. Tutorials of Cytology, Chicago, 1974, pp. 40–53.
3a. Hajdu, S. I. and Melamed, M. R.: The diagnostic value of aspiration smears. *Am J Clin Pathol, 59:* 350–356, 1973.

Cellular Changes Associated With Nonneoplastic Diseases

CELLS OF EPITHELIAL ORIGIN

As the production of sputum in quantities sufficient to be expectorated does not occur in the absence of respiratory disease, the examination of such specimens generally demands a greater challenge than that encountered in the screening of genital smears from asymptomatic women. Although sputum production in itself is an abnormal phenomenon, its cellular components may be unremarkable. The basic morphology of these cellular components has been well described in the literature by Farber and his associates,[1] Woolner and McDonald,[2–9] Koss,[10] and more recently by Frost and his associates.[11] The normal epithelial components of sputum consist of squamous epithelial cells exfoliating from the oral cavity and pharynx, respiratory columnar epithelium exfoliating most frequently from the tracheobronchial tree and less frequently from the upper respiratory passages, terminal bronchiolar epithelium, and alveolar pneumocytes.

Normal Bronchial Columnar Epithelial Cells

The tall columnar cells lining the tracheobronchial tree may be of two varieties. Most frequently encountered are ciliated columnar cells. They are most commonly seen in bronchial washings, aspirations, or brushings. They should not be present in large numbers in sputum except in postbronchoscopy specimens. When many columnar cells are present in routine specimens of sputum in the absence of prior instrumentation, disease involving the respiratory epithelium should be suspected. Among the many causative factors may be noted bacteria, viruses, and miscellaneous irritants. Figures III-1 through III-8 show examples of ciliated bronchial columnar cells as they present in samples of sputum and bronchial material. They are characterized in profile by a columnar or prismatic shape ending in a tail. The nucleus is oriented toward the tail end and shows a finely granular chromatin pattern with one or more small nucleoli. Cilia with a terminal plate are present. During degeneration the cilia are often lost, leaving the cell with just a terminal plate.

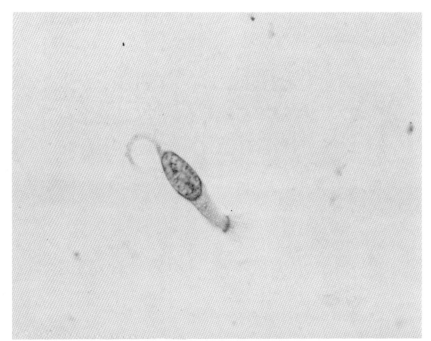

Fig. III-1. Bronchial washing. Membrane filter preparation. Normal ciliated bronchial columnar cell. The nuclear chromatin is finely granular and uniformly distributed throughout the nucleus. Note the closeness of the cytoplasmic membrane and the nuclear membrane in the lateral portions of the cell. A cytoplasmic tail is visible in the 11 o'clock position. Cilia and a terminal plate are visible at the 4 o'clock position. Papanicolaou stain, × 1000.

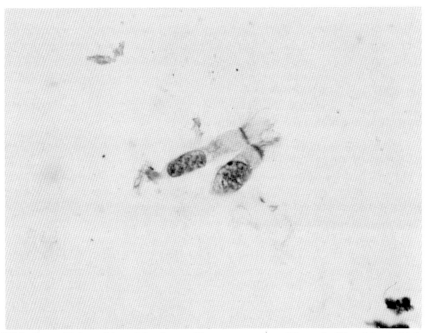

Fig. III-2. Bronchial washing. Membrane filter preparation. Normal ciliated bronchial columnar cells. Papanicolaou stain, × 1000.

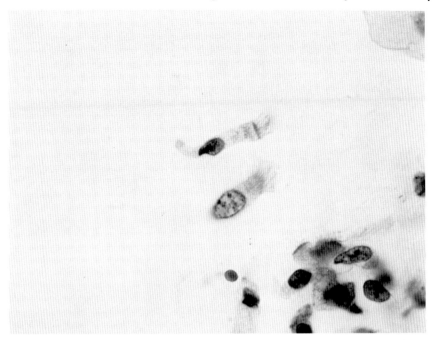

Fig. III-3. Bronchial brushing. Membrane filter preparation. Normal cilitated bronchial columnar cells. Papanicolaou stain, × 1000.

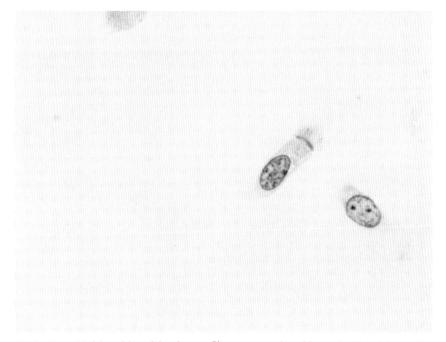

Fig. III-4. Bronchial brushing. Membrane filter preparation. Normal ciliated bronchial columnar cells. Papanicolaou stain, × 1000.

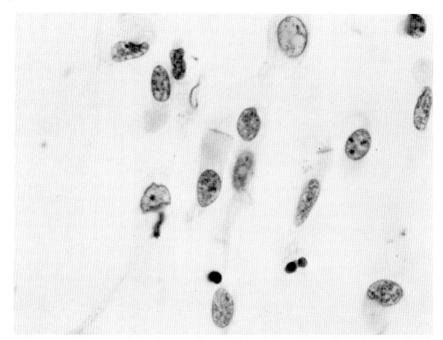

Fig. III-5. Fresh sputum. Normal ciliated bronchial columnar cells. Note the variation in size and shape of the nuclei in these cells, which otherwise are unremarkable. Papanicolaou stain, × 1000.

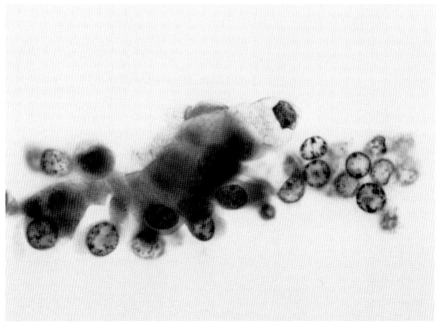

Fig. III-6. Bronchial washing. Membrane filter preparation. Normal ciliated bronchial columnar cells and mucus-producing columnar cells. Papanicolaou stain, × 1000.

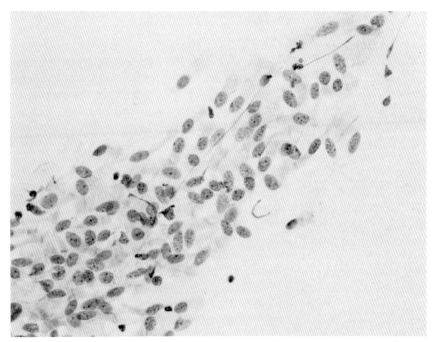

Fig. III-7. Fresh sputum. Normal ciliated bronchial columnar cells. Papanicolaou stain, ×
1000.

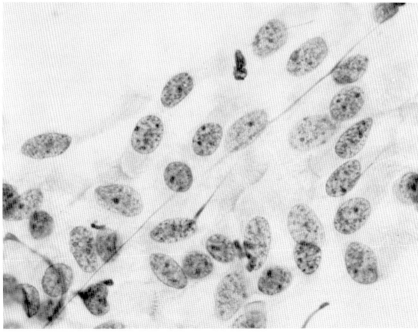

Fig. III-8. Fresh sputum. Normal ciliated bronchial columnar cells as observed in Figure III-
7 at higher power. Many of the nuclei in the cells lying in profile appear to extend far beyond
the boundaries of the cytoplasm. Note in particular the cell at the 9 o'clock position. Note also
the frequent tendency of nuclei to assume a comma shape in conformity with the cytoplasmic
tail. These are commonly encountered variants of normal bronchial cell morphology. Papanicolaou stain, × 1000.

Encountered less frequently are mucus-producing bronchial cells, so-called goblet cells. Multiple examples of these are seen in Figures III-9 through III-13. In profile a large hypersecretory vacuole, which distends the cytoplasm and distorts the nuclear shape, can be observed in that portion of the cytoplasm between the nucleus and the luminal portion of the cell. These cells are more common in patients with chronic tracheobronchial disease such as chronic bronchitis, asthmatic bronchitis, and bronchiectasis.

Abnormal Bronchial Columnar Epithelial Cells

So-called irritation forms or reactive forms of bronchial epithelium may occur in response to a host of insults, varying from microbiological pathogens to environmental irritants.[11-21] Figures III-14 through III-29 exhibit examples of such cells. The cells may become markedly enlarged with a coarsening of the nuclear chromatin pattern and the appearance of one or more enlarged nucleoli. The presence of a terminal plate with cilia, although quite degenerate, may be of aid in assuring the examiner that the cell is benign. Note that tissue fragments as well as single cells may occur. The cellular patterns illustrated most likely represent a spectrum of individual reaction ranging from one of mild enlargement with slight nuclear atypias to very bizarre changes perhaps best seen in cellular alterations in response to chemotherapy. Components of hyperplasia, regeneration, and metaplasia may further complicate

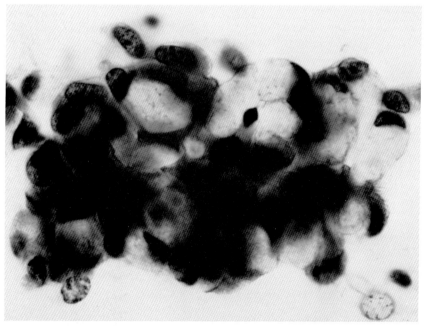

Fig. III-9. Bronchial washing. Membrane filter preparation. Bronchial epithelial cells exhibiting mucus production. Such cells are seen only occasionally in the usual specimen of sputum or bronchial washing. The most striking feature of such cells is the presence of large, single, or occasionally multiple hyperdistended secretory-type vacuoles, which markedly distend the outline of the cells and produce lateral displacement and molding of the nuclei. Papanicolaou stain, × 1000.

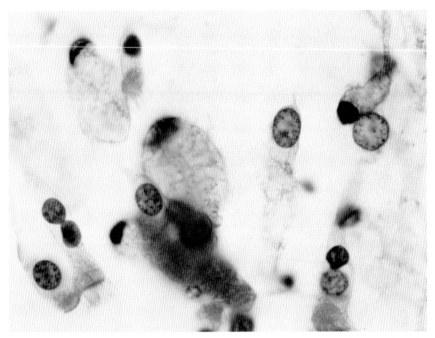

Fig. III-10. Bronchial washing. Membrane filter preparation. Same specimen as that depicted in Figure III-9. Another field of mucus-producing columnar cells is shown. Note the marked distortion of the nucleus of the cell at the 11 o'clock position. Papanicolaou stain, × 1000.

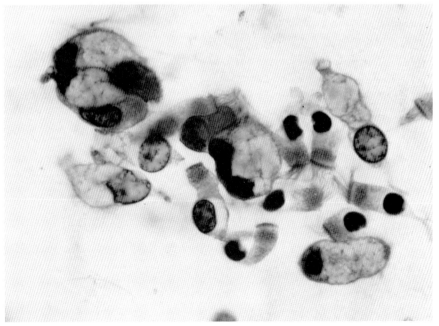

Fig. III-11. Bronchial washing. Membrane filter preparation. Mucus-producing bronchial columnar cells. Same case as Figures III-9 and III-10. Papanicolaou stain, × 1000.

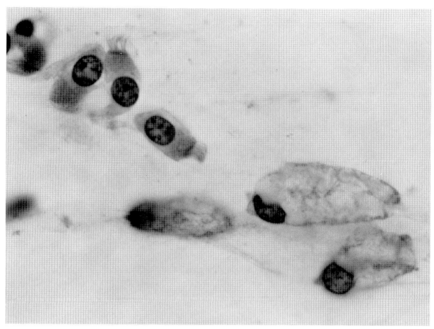

Fig. III-12. Bronchial washing. Membrane filter preparation. Mucus-producing bronchial columnar cells. Same case as Figures III-9–III-11. Papanicolaou stain, × 1000.

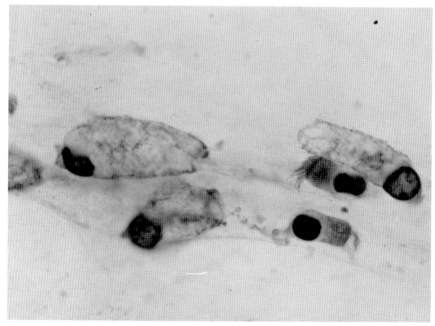

Fig. III-13. Bronchial washing. Membrane filter preparation. Mucus-producing bronchial columnar cells. From same case as Figures III-9–III-12. Papanicolaou stain, × 1000.

Fig. III-14. Fresh sputum. Irritation form of ciliated bronchial columnar cell. The entire cell is enlarged, with increases in size of both nucleus and cytoplasm. The nucleus exhibits an increase in the granularity of the chromatin pattern with hyperchromasia. One prominent nucleolus is present. The remains of degenerating cilia may be seen with a prominently staining terminal plate. Papanicolaou stain, × 1000.

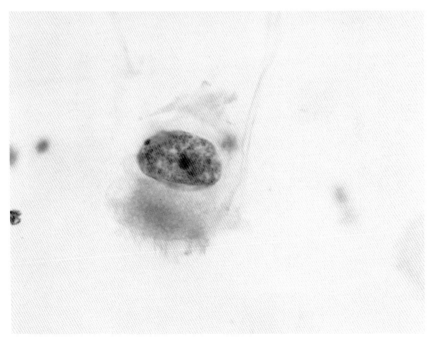

Fig. III-15. Fresh sputum. Irritation form of ciliated bronchial columnar epithelial cell. Compare this cell with that shown in Figure III-14 at the same magnification. This cell is even larger and shows more strikingly the features described for the previous cell. This cell is in an even more advanced state of degeneration, with disappearance of the cilia and fragmentation of the cytoplasm. Papanicolaou stain, × 1000.

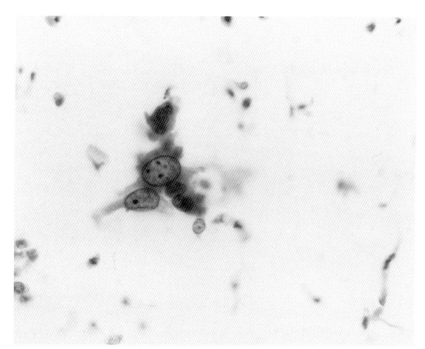

Fig. III-16. Bronchial brushing. Membrane filter preparation. Irritation forms of bronchial columnar epithelium. Papanicolaou stain, × 400.

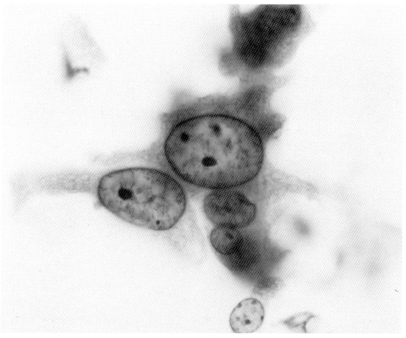

Fig. III-17. Bronchial brushing. Membrane filter preparation. These are the cells depicted in Figure III-16, shown at higher magnification. The nuclei are irregular in size and shape and exhibit enlarged nucleoli. Nucleoli are multiple in at least one cell. Some chromocenters are present, but intense granularity of chromatin is not present in these cells. Although the architecture is altered the cells remain tightly cohesive. Papanicolaou stain, × 1000.

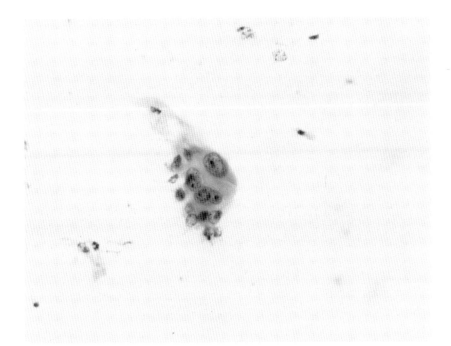

Fig. III-18. Bronchial brushing. Membrane filter preparation. Irritation forms of bronchial columnar cells. Papanicolaou stain, × 400.

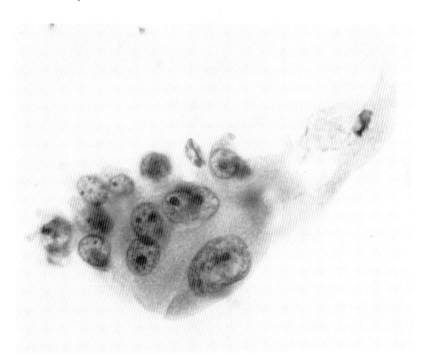

Fig. III-19. Bronchial brushing. Membrane filter preparation. Irritation forms of bronchial columnar cells. This is a higher magnification of the cells shown in Figure III-18. These cells exhibit both hyperchromasia and prominent multiple nucleoli in the nuclei. The nuclei show marked variation in size and shape. Although these are features which would give one concern about the possibility of malignancy, the tightly cohesive relationship between the cells as well as the cellular monolayer are features against malignancy. Papanicolaou stain, × 1000.

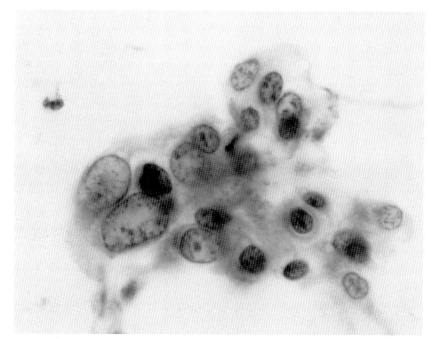

Fig. III-20. Bronchial washing. Membrane filter preparation. Reaction forms of bronchial columnar epithelium. Papanicolaou stain, × 1000.

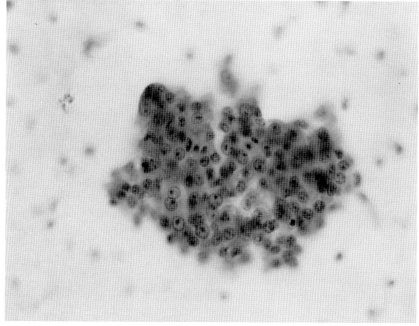

Fig. III-21. Bronchial washing. Membrane filter preparation. Reaction forms of bronchial columnar epithelium. Papanicolaou stain, × 400.

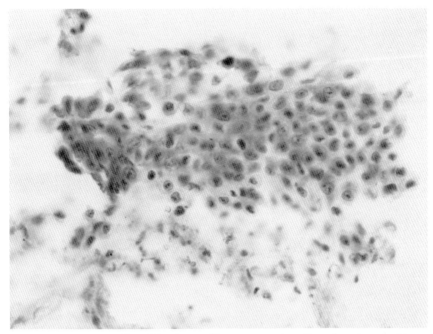

Fig. III-22. Bronchial washing. Membrane filter preparation. Reaction forms of bronchial columnar epithelium. Papanicolaou stain, × 400.

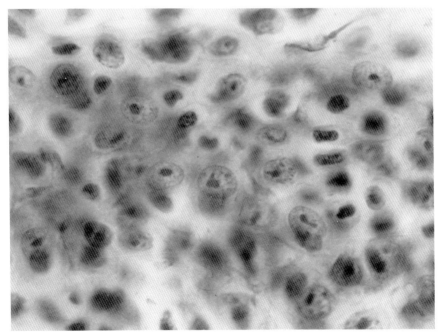

Fig. III-23. Bronchial washing. Membrane filter preparation. The same microscopic field as depicted in Figure III-22 at a higher power. The enlarged irregular nucleoli are especially striking in this field. The cells, however, are tightly cohesive. Papanicolaou stain, × 100.

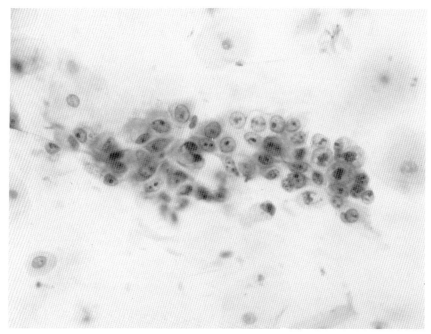

Fig. III-24. Fresh sputum. Reaction forms of bronchial columnar epithelium. Papanıcolaou stain, × 400.

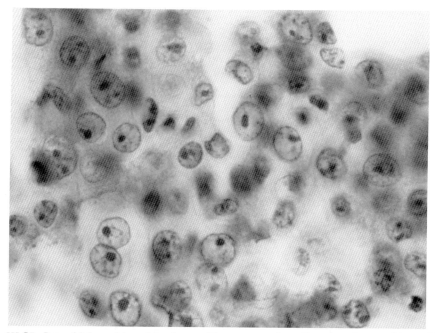

Fig. III-25. Bronchial washing. Membrane filter preparation. Reaction forms of bronchial columnar epithelium. Papanicolaou stain, × 1000.

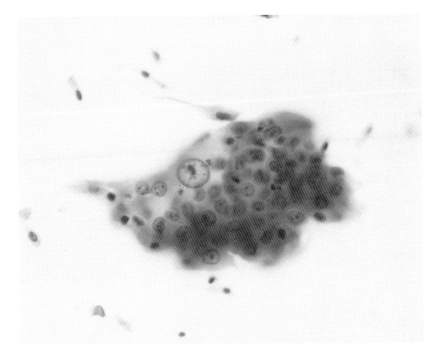

Fig. III-26. Bronchial brushing. Membrane filter preparation. Reaction forms of bronchial columnar epithelium. Papanicolaou stain, × 400.

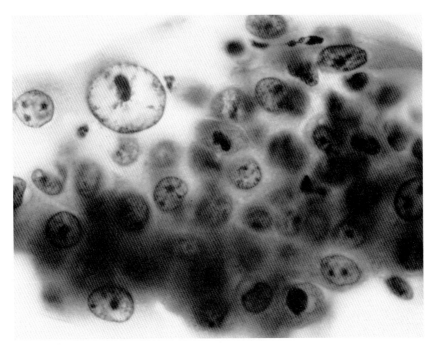

Fig. III-27. Bronchial brushing. Membrane filter preparation. Reaction forms of bronchial columnar epithelium as shown in Figure III-26 at higher magnification. In Figures III-26 and III-27, in the left upper portion of the photographs, a flattening suggestive of a luminal border is present. Note the irregularity in size and shape of the nuclei, with an unusually prominent nucleolus depicted at the 11 o'clock position in Figure III-27. Papanicolaou stain, × 1000.

33

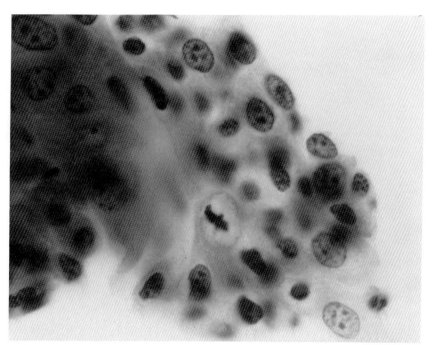

Fig. III-28. Bronchial brushing. Membrane filter preparation. Reaction forms of bronchial columnar epithelium. Another field from the specimen depicted in Figures III-26 and III-27. Papanicolaou stain, × 1000.

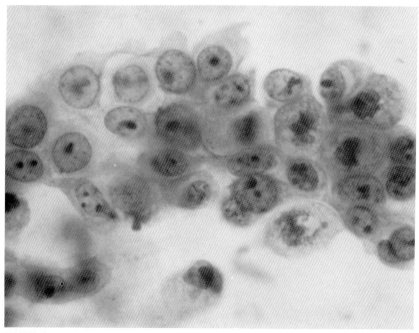

Fig. III-29. Fresh sputum. Reaction forms of bronchial columnar epithelium. Papanicolaou stain, × 1000.

the picture. Extreme care must be exercised in the presence of such alterations to avoid an erroneous diagnosis of cancer. The authors believe that these abnormal cells and tissue fragments have become more frequently observed in the years since the introduction of bronchial brushing techniques. A major characteristic of the benign tissue fragments in these specimens is the preservation of tight cohesion between cells, a characteristic frequently absent from neoplastic tissue fragments.

Another common response to irritation is multinucleation.[22, 23] Figures III-30 through III-38 illustrate varying manifestations of multinucleated bronchial columnar cells. Nuclei are small and mirror-image. Although the appearance of such cells may follow a wide variety of insults, including chemical agents, infection, and ionizing radiation, they are most commonly seen following instrumentation, particularly bronchoscopy.

Hyperplasia of mature bronchial cells may occur in response to a number of chronic diseases of the lung, particularly bronchiectasis,[24] chronic bronchitis, and asthma.[25, 26] It was in patients with chronic asthmatic bronchitis that the papillary tissue fragments exfoliating from such hyperplasia were first noted and mistakenly diagnosed as adenocarcinoma. Such fragments of hyperplastic cells were given the name "Creola" bodies as a reminder of the first patient in whom a mistaken diagnosis of adenocarcinoma was made. Indeed, these fragments are said to appear in the sputum from 42% of cases of asthmatic bronchitis. Figures III-39 through III-52 show examples of bronchial hyperplasia. The cytologic presentation is that of papillary ciliated respiratory epithelium. There is some nuclear molding between individual cells, although intranuclear chromatin and nucleolar structures remain relatively unremarkable. A varying number of highly vacuolated mucus-producing cells may also be present in these fragments. It is extremely important that they be accurately recognized and not be mistaken for adenocarcinoma. The key to their benignancy is to be found in the finely granular chromatin pattern, regular uniform nucleoli, and the presence of cilia.

Cellular Changes Originating in Terminal Bronchiolar and Alveolar Epithelium

The least frequently encountered epithelial components in sputum are cells exfoliating from the terminal bronchioles and from the alveoli. Considerable attention has been given in the literature to the association of these cells with such respiratory problems as pulmonary fibrosis, thermal injury, thromboembolism with or without pulmonary infarction, anthracosis, and chronic organizing pneumonia. The relationship of these epithelial changes to the development of bronchiolo-alveolar cell carcinoma has been the subject of much interest, and the presence of these cells in sputum may represent a particularly difficult problem in interpretation to the cytopathologist.[27-32]

Although the utilization of such relatively modern techniques as transmission and scanning electron microscopy and biochemical techniques have enabled us to differentiate a variety of subtypes of terminal bronchiolar and alveolar cells, conventional light microscopy of cytologic specimens does not permit the observer to appreciate these various cell types. Practically speaking the terminal bronchiolar and alveolar cells are probably not recognized in sputum when they are present in an unaltered

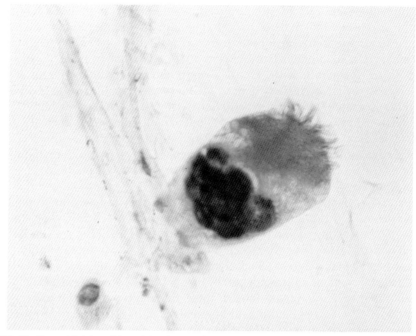

Fig. III-30. Bronchial brushing. Membrane filter preparation. Multinucleated ciliated bronchial columnar cell. In this cell the blurring of the nuclei is secondary to their degeneration. Papanicolaou stain, × 1000.

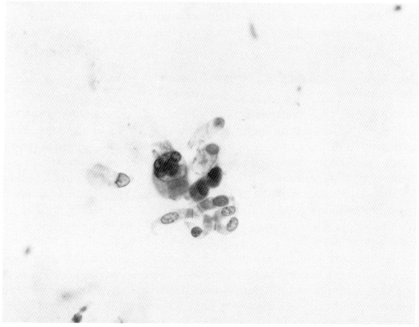

Fig. III-31. Bronchial washing. Membrane filter preparation. Multinucleated ciliated bronchial columnar cell and normal ciliated bronchial columnar cells. Papanicolaou stain, × 400.

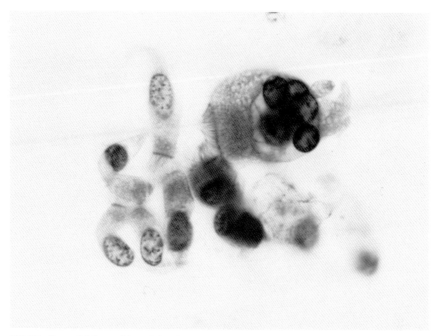

Fig. III-32. Bronchial washing. Membrane filter preparation. Multinucleated ciliated bronchial columnar cell and normal ciliated bronchial columnar cells as depicted in Figure III-31 at higher magnification. In the multinucleated cell, the nuclei are well preserved and are mirror image in type. Papanicolaou stain, × 1000.

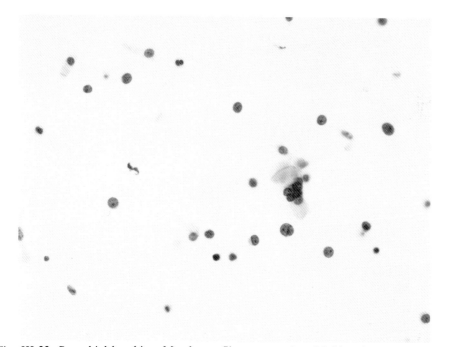

Fig. III-33. Bronchial brushing. Membrane filter preparation. Multinucleated ciliated bronchial columnar cell. Papanicolaou stain, × 400.

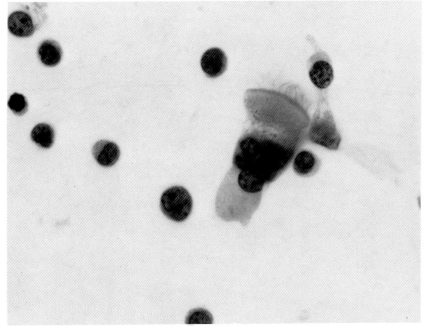

Fig. III-34. Bronchial brushing. Membrane filter preparation. Multinucleated ciliated bronchial columnar cell as shown in Figure III-33 at higher magnification. Papanicolaou stain, × 1000.

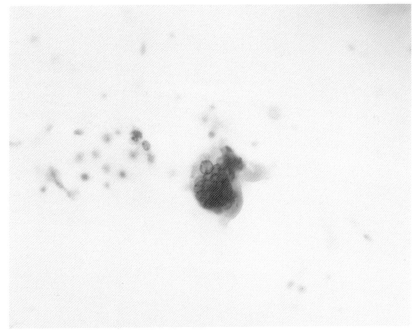

Fig. III-35. Bronchial washing. Membrane filter preparation. Multinucleated ciliated bronchial columnar cell. Papanicolaou stain, × 400.

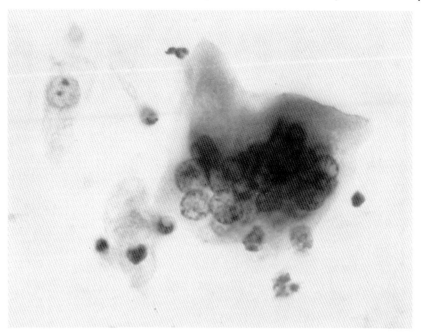

Fig. III-36. Bronchial washing. Membrane filter preparation. Multinucleated ciliated bronchial columnar epithelial cell, as shown in Figure III-35, but at higher magnification. Papanicolaou stain, × 1000.

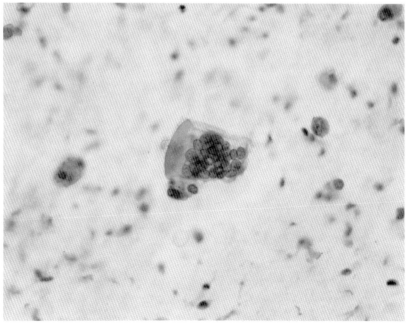

Fig. III-37. Bronchial washing. Membrane filter preparation. Multinucleated ciliated bronchial columnar cell. Papanicolaou stain, × 400.

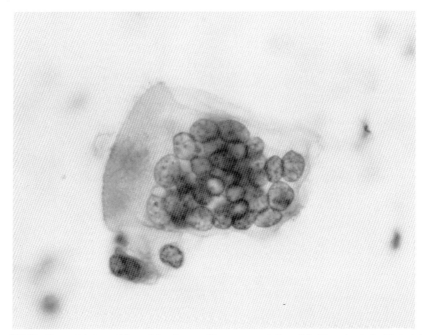

Fig. III-38. Bronchial washing. Membrane filter preparation. Multinucleated ciliated bronchial columnar cell, as shown in Figure III-37, at higher power. This photograph illustrates the large number of nuclei which may be formed in the cell. Papanicolaou stain, × 1000.

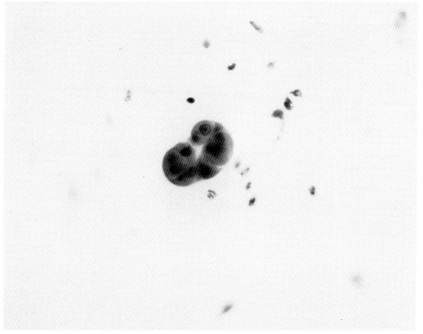

Fig. III-39. Fresh sputum. Hyperplasia of bronchial columnar epithelium. Papanicolaou stain, × 400.

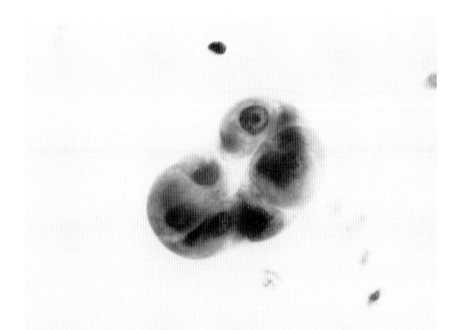

Fig. III-40. Fresh sputum. Hyperplasia of bronchial columnar epithelium. Some of the characteristic features of hyperplasia are illustrated in this fragment, which is at a higher magnification than that depicted in Figure III-39. Upon exfoliation there is a tendency for the fragment to form itself into a ball-like cluster of cells such that at times it seems to be surrounded on all sides by a luminal border. Individual cells show enlargement both of nuclei and cytoplasm. Nucleoli may be quite prominent. There may be intense molding between individual cells. Cytoplasm may vary from homogenous or foamy to the containment of multiple large vacuoles. Papanicolaou stain, × 1000.

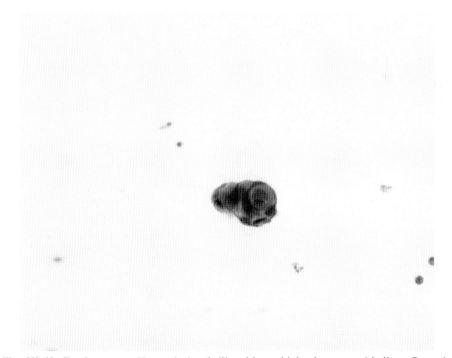

Fig. III-41. Fresh sputum. Hyperplasia of ciliated bronchial columnar epithelium. Papanicolaou stain, × 400.

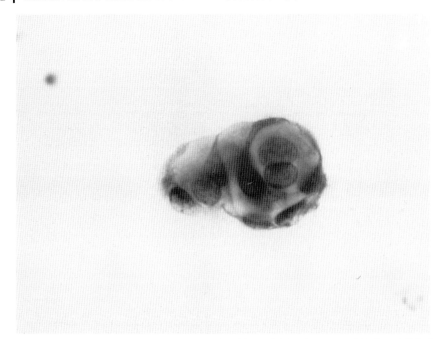

Fig. III-42. Fresh sputum. Hyperplasia of ciliated bronchial columnar epithelium. This figure shows a higher magnification of the cellular cluster depicted in Figure III-41. Cilia in varying degrees of degeneration are visible around the entire circumference of the fragment. Papanicolaou stain, × 1000.

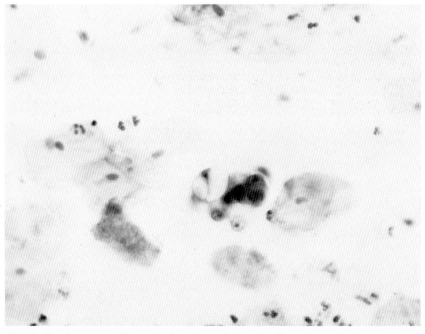

Fig. III-43. Fresh sputum. Hyperplasia of bronchial epithelium. Papanicolaou stain, × 400.

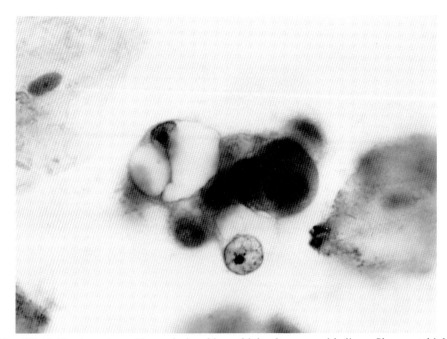

Fig. III-44. Fresh sputum. Hyperplasia of bronchial columnar epithelium. Shown at higher magnification is the cell cluster depicted in Figure III-43. The feature of marked hyperdistended secretory vacuoles as an occasional component of hyperplasia is shown here. Enlarged nucleoli are seen in several of the nuclei. Confusion with bronchogenic adenocarcinoma is very easy with cellular fragments such as these. A faintly visible border of cilia present along the top border of the cellular cluster is a major aid in preventing false cancer diagnosis from this cluster. Papanicolaou stain, × 1000.

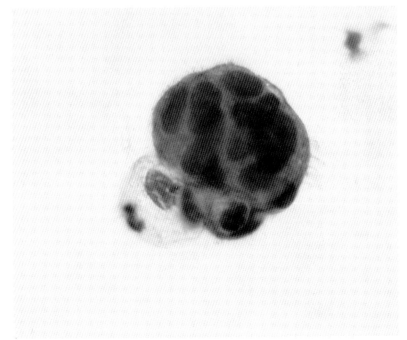

Fig. III-45. Fresh sputum. Hyperplasia of bronchial columnar epithelium. Papanicolaou stain, × 1000.

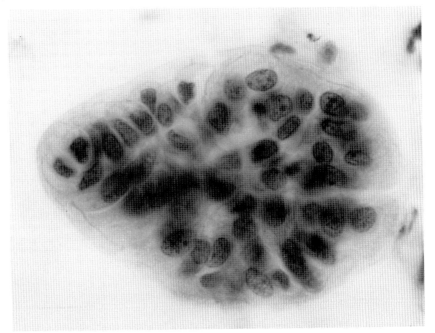

Fig. III-50. Fresh sputum. Hyperplasia of bronchial columnar epithelium. This is a higher magnification of the cell cluster shown in Figure III-49. Papanicolaou stain, × 1000.

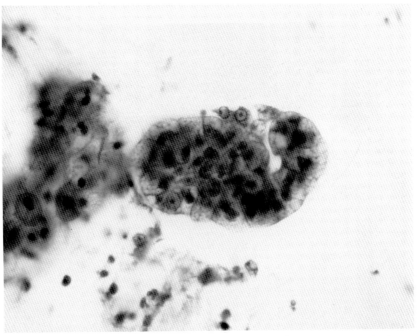

Fig. III-51. Fresh sputum. Hyperplasia of bronchial columnar epithelium. Papanicolaou stain, × 400.

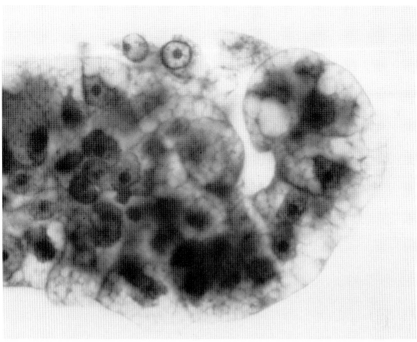

Fig. III-52. Fresh sputum. Hyperplasia of bronchial columnar epithelium. This is a higher magnification of the tissue fragment shown in Figure III-51. There may be confusion between this benign tissue fragment and one resulting from mucus-producing papillary adenocarcinoma. Papanicolaou stain, × 1000.

form. These cells are relatively small and when present in sputum appear as rounded single cells with finely vacuolated cytoplasm and centrally placed nuclei without abnormalities. As such they are usually mistaken for alveolar macrophages (Figs. III-53 and III-54). Only in the presence of insult do they enlarge and become problematical. In such circumstances they may be formed together as small papillary tissue fragments, individual cells of which show enlarged, centrally placed nuclei with one or more stainable nucleoli (Figs. III-55 through III-59). The cytoplasm may be granular and finely vacuolated, or may exhibit one or more hyperdistended vacuoles (Figs. III-60 through III-64). Differential diagnosis of such cells then becomes a rather formidable problem of determining whether these cells are coming from some of the above-named benign disease processes or whether they are derived from a bronchiolo-alveolar cell carcinoma. Pulmonary infarcts in particular may give rise to cells of this description, and to the experienced cytopathologist are known to be one of the most dangerous sources for diagnostic error. These differential diagnostic problems are discussed further in Chapter VIII.

Reserve-Cell Hyperplasia and Squamous Metaplasia

The concept of squamous metaplasia of the respiratory mucosa has been applied to a spectrum of alterations, beginning with reserve-cell hyperplasia and ending with a stratified and keratinized covering resembling mature squamous epithelium. Its

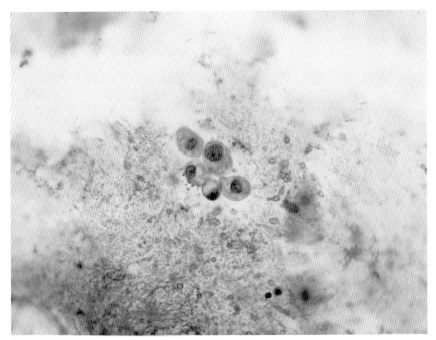

Fig. III-53. Fresh sputum. Cells believed to be of bronchiolor-alveolar epithelial origin. Papanicolaou stain, × 400.

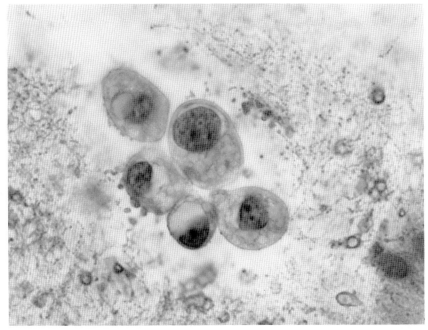

Fig. III-54. Fresh sputum. Cells believed to be of bronchiolor-alveolar cell origin. These cells are characterized by a shape that varies from ovoid to polygonal. The nuclei may be either centrally placed or eccentrically placed so that they resemble macrophages. Nuclear chromatin is finely granular with one or more visible nucleoli. The cytoplasm may exhibit varying degrees of vacuolization. These cells, when seen as single cells, are usually interpreted as alveolar macrophages. Papanicolaou stain, × 1000.

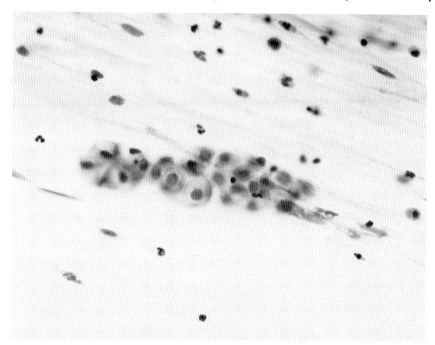

Fig. III-55. Fresh sputum. Cells believed to originate from bronchiolor-alveolar epithelium. Papanicolaou stain, × 400.

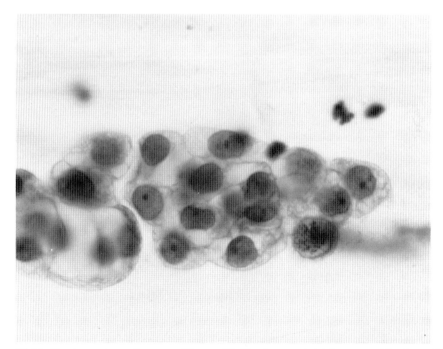

Fig. III-56. Fresh sputum. Same as Figure III-55, shown at higher magnification. Cells believed to be derived from bronchiolor-alveolar epithelium. Papanicolaou stain, × 1000.

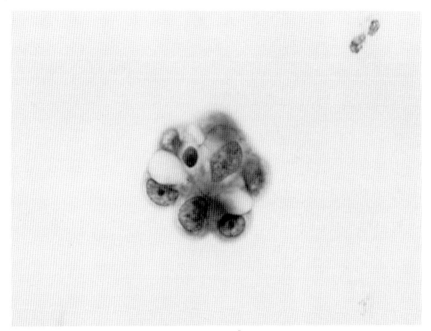

Fig. III-57. Fresh sputum. Cells believed to be derived from bronchiolor-alveolar epithelium. Hyperdistended secretory vacuoles are numerous and prominent in this cluster of cells. Papanicolaou stain, × 1000.

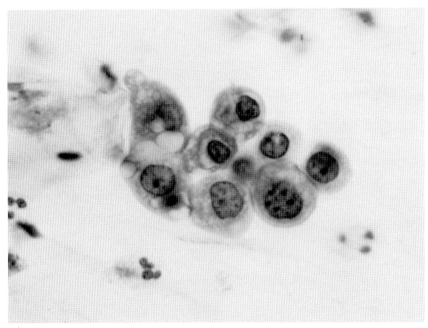

Fig. III-58. Fresh sputum. Cells believed to have originated in bronchiolor-alveolar epithelium. Same specimen but different microscopic field from that shown in Figure III-57. Papanicolaou stain, × 1000.

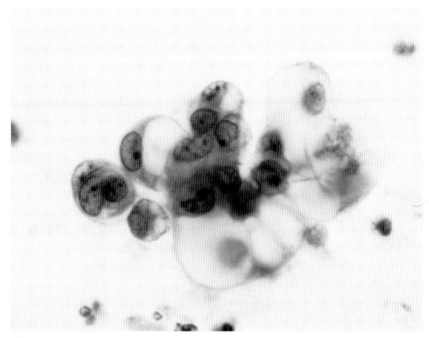

Fig. III-59. Fresh sputum. Cells believed to have originated in bronchiolor-alveolar epithelium. Secretory vacuoles are very striking in this tissue fragment. There is a moderate variation in nuclear size and shape, with many prominent nucleoli. Papanicolaou stain, × 1000.

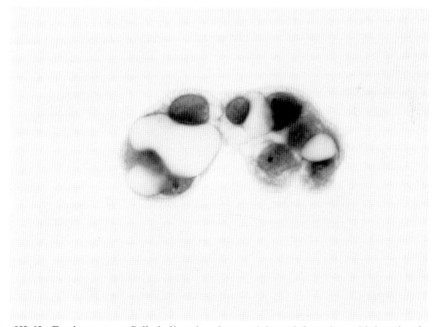

Fig. III-60. Fresh sputum. Cells believed to have originated from bronchiolor-alveolar epithelium. Papanicolaou stain, × 1000.

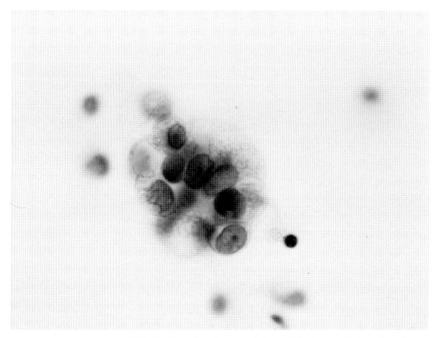

Fig. III-61. Fresh sputum. Cells believed to have originated in bronchiolor-alveolar epithelium. Papanicolaou stain, × 1000.

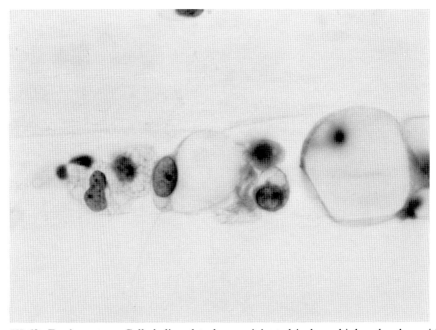

Fig. III-62. Fresh sputum. Cells believed to have originated in bronchiolor-alveolar epithelium. Another field from the specimen photographed in Figure III-61. Papanicolaou stain, × 1000.

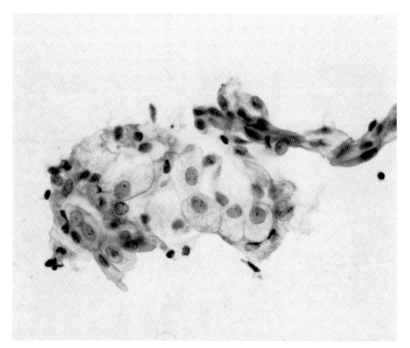

Fig. III-63. Thin-needle aspiration biopsy of the lung. Membrane filter preparation. Cells believed to have originated in bronchiolor-alveolar epithelium. Papanicolaou stain, × 400.

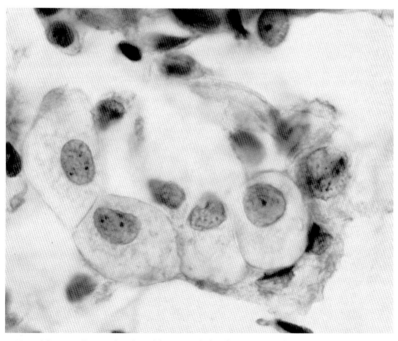

Fig. III-64. Thin-needle aspiration biopsy of the lung. Membrane filter preparation. Cells believed to have originated in bronchiolor-alveolar epithelium. A higher magnification of the group shown in Figure III-63. Note the chain-like configuration of these plump epithelial cells, with some suggestion that they were surrounding a lumen or cavity. Papanicolaou stain, × 1000.

occurrence in individuals exposed to varying environmental toxic agents, particularly cigarette smokers, and its possible relationship to the pathogenesis of bronchogenic carcinoma have become increasingly interesting to investigators.[10, 11, 13–21, 33–42]

In association with alterations of the covering columnar epithelial cells previously noted, there begins proliferation of reserve cells such that a multilayered epithelium, so-called reserve cell hyperplasia, is produced, which intervenes between the overlying columnar epithelium and the basement membrane. Exfoliation of the columnar cell layer eventually occurs, leaving behind an epithelium composed of immature reserve cells. As they in turn gradually mature, an epithelium is produced which more and more resembles a squamous type, with cell flattening, karyopyknosis, and keratin production. Reserve-cell hyperplasia in cytologic materials is recognized by the presence of tissue fragments composed of small, uniform, tightly coherent cells possessing darkly stained nuclei and a thin rim of faintly cyanophilic cytoplasm (Figs. III-65 through III-70). There is nuclear molding, but uniformity exists throughout the fragment. There is no tendency toward fragmentation of the cluster. Toward the edge one may see some maturation and more columnar configuration. There should be no necrosis or *tumor diathesis* in the background. At times these cells may be alarming in appearance and must be distinguished from small-cell anaplastic carcinoma.[43] Other small-cell neoplasms, notably leukemias and lymphomas, should not be confused with reserve-cell hyperplasia, as they characteristically shed into the broncho-pulmonary material as single cells.

Cells derived from squamous metaplasia are shown in Figures III-71 through III-

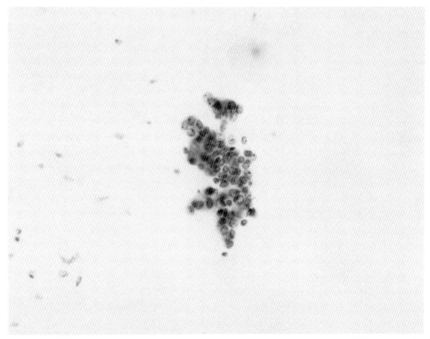

Fig. III-65. Bronchial washing. Membrane filter preparation. Reserve-cell hyperplasia. Papanicolaou stain, × 400.

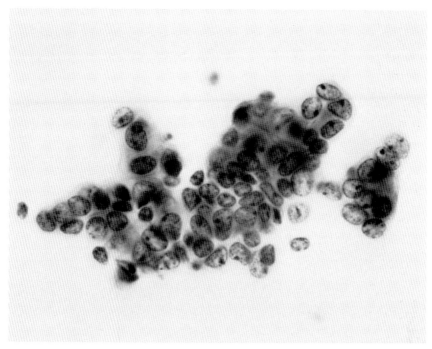

Fig. III-66. Bronchial washing. Membrane filter preparation. Reserve-cell hyperplasia. This is a higher magnification of the group shown in Figure III-65. The characteristic features of reserve-cell hyperplasia are strikingly shown in this group. The nuclei are small and fairly uniform. They are tightly coherent. The small amount of cytoplasm present between the nuclei usually stains intensely cyanophilically with the Papanicolaou stain. The nuclei may vary markedly in their degree of preservation. Some show a vesicular nucleus with finely granular chromatin and an occasional small nuclei. Other nuclei may exhibit only markedly degenerated contents. There should be no evidence of necrosis in the background or a tendency of the cells to pull away from one another. Papanicolaou stain, × 1000.

87. They may be seen as single cells and as small tissue fragments. In the latter the cells are grouped in uniform, monolayered cobblestoned-like arrangement with striking uniformity between the cells. Although they resemble mature squamous cells, they are smaller and possess a higher nucleus to cytoplasmic ratio. As squamous metaplasia mimics maturing squamous epithelium, metaplastic cells of varying degrees of maturity may be present. The cytoplasmic staining characteristics may vary from a deep cyanophilia to an orangeophilia, indicating maturation and keratinization of the cytoplasm. The nuclei may be intensely karyopyknotic (Figs. III-86 and III-87). The nuclei in squamous metaplasia are characterized by roundness, uniformity, and a tendency toward karyopyknosis in the more mature varieties. However, occasional examples will show nuclear abnormalities of increasing degrees of severity. These abnormalities are characterized by an increase in the nuclear cytoplasmic ratio, thickening of the nuclear membrane, increasing granularity and hyperchromasia of the chromatin, and the appearance of nucleoli. In one of the authors' (Johnston) laboratories this change has been called *squamous metaplasia with focal dysplasia.* These changes have been observed in the presence of long-

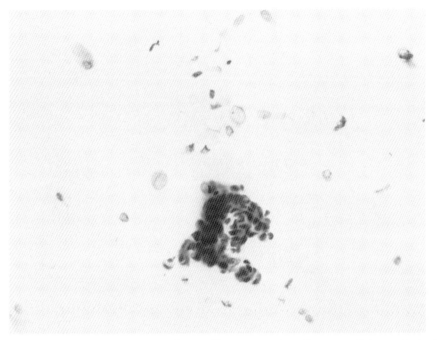

Fig. III-67. Bronchial washing. Membrane filter preparation. Reserve-cell hyperplasia. Papanicolaou stain, × 400.

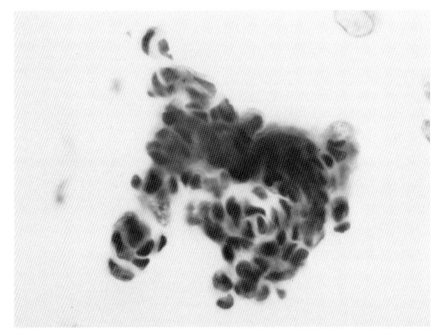

Fig. III-68. Bronchial washing. Membrane filter preparation. Reserve-cell hyperplasia. The same cellular group as shown in Figure III-67. In this tissue fragment all of the nuclei show advanced degeneration. This is a frequent finding. Papanicolaou stain, × 1000.

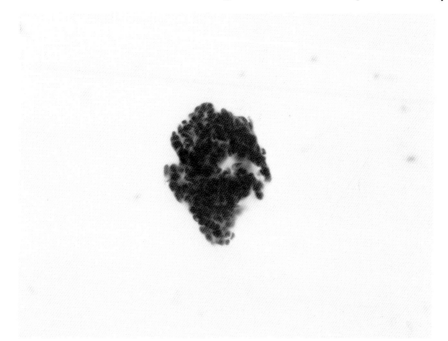

Fig. III-69. Bronchial washing. Membrane filter preparation. Reserve-cell hyperplasia. Papanicolaou stain, × 400.

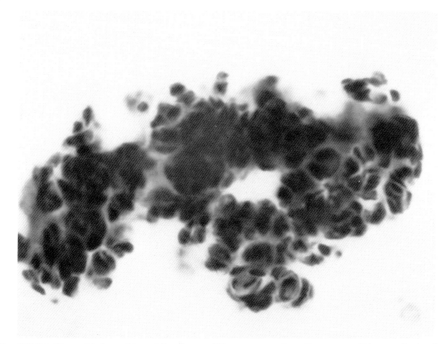

Fig. III-70. Bronchial washing. Membrane filter preparation. Reserve-cell hyperplasia. Same tissue fragment at higher magnification as shown in Figure III-69. There is marked nuclear degeneration and striking molding between cells. However, the clusters are tightly cohesive and there is no necrosis in the background. Papanicolaou stain, × 1000.

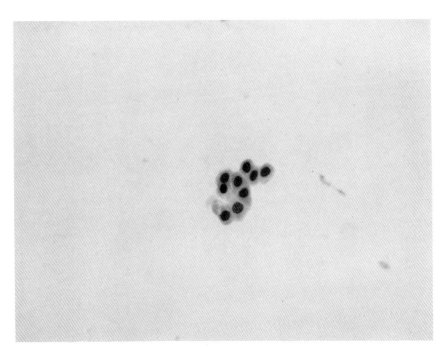

Fig. III-71. Fresh sputum. Squamous metaplasia. Papanicolaou stain, × 400.

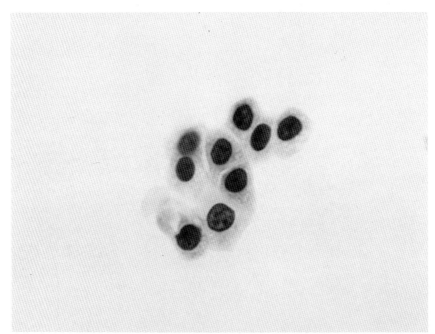

Fig. III-72. Fresh sputum. Squamous metaplasia. Higher magnification of the tissue fragment shown in Figure III-71. These two figures show the most commonly observed cytologic manifestation of squamous metaplasia. The tissue fragment consists of a monolayer of small cells arranged in a pattern similar to a mosaic. The cells resemble squamous cells, but the nuclear cytoplasmic ratios are too great for normal squamous cells. The nuclei exhibit varying degrees of degeneration, some already in beginning karyopyknosis. Papanicolaou stain, × 1000.

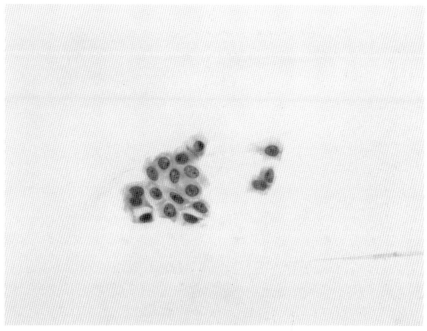

Fig. III-73. Fresh sputum. Squamous metaplasia. Papanicolaou stain, × 400.

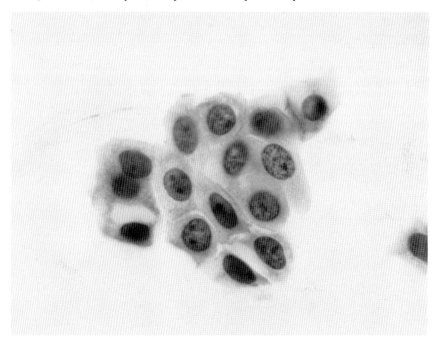

Fig. III-74. Fresh sputum. Squamous metaplasia. A higher magnification of the tissue fragment shown in Figure III-73. This fragment of metaplastic epithelium exhibits more maturity than that shown in Figures III-71 and III-72. These cells have a larger amount of cytoplasm and there is more flattening of the cytoplasm. The cleavage between the cells is very marked. Nuclear chromatin is finely granular in nuclei, which are preserved. An occasional nucleolus may be present. Papanicolaou stain, × 1000.

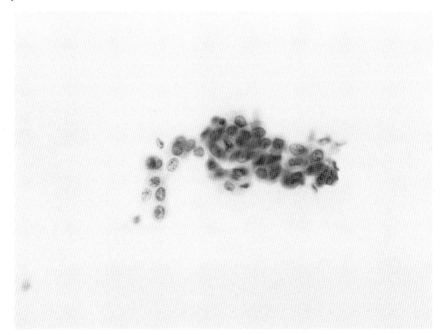

Fig. III-75. Bronchial brushing. Membrane filter preparation. Squamous metaplasia. Papanicolaou stain, × 400.

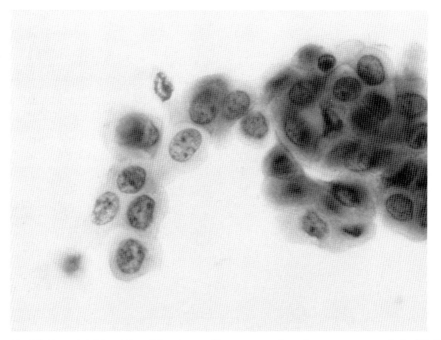

Fig. III-76. Bronchial brushing. Membrane filter preparation. Squamous metaplasia. Higher magnification of the field shown in Figure III-75. Papanicolaou stain, × 1000.

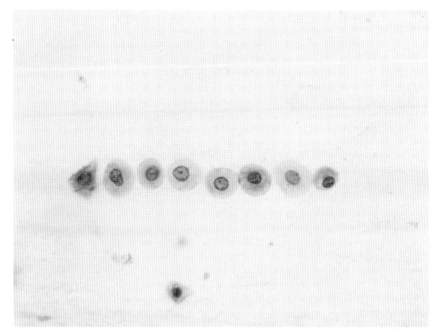

Fig. III-77. Fresh sputum. Squamous metaplasia. Papanicolaou stain, × 400.

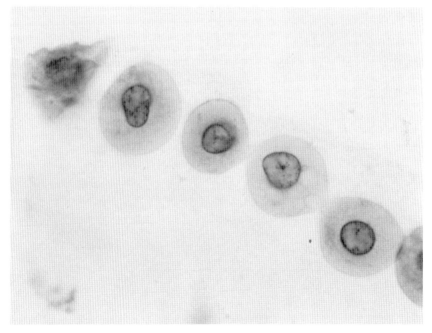

Fig. III-78. Fresh sputum. Squamous metaplasia. Higher magnification of the cells shown in Figure III-77. Papanicolaou stain, × 1000.

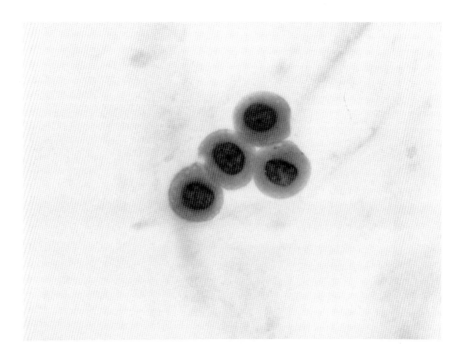

Fig. III-79. Fresh sputum. Squamous metaplasia with dysplasia. There is beginning evidence of nuclear abnormality, characterized by an increase in the nuclear cytoplasmic ratio, a thickening of the nuclear membrane, and an increase in the granularity and hyperchromasia of nuclear chromatin. Several small nucleoli are present. Papanicolaou stain, × 1000.

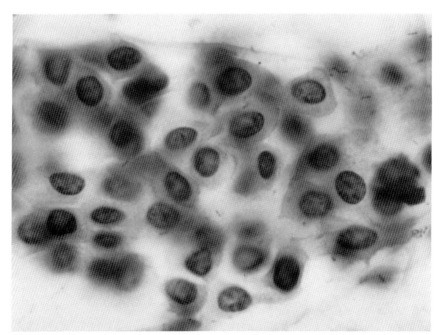

Fig. III-80. Fresh sputum. Squamous metaplasia with dysplasia. In this large tissue fragment a beginning distortion of architectural arrangement can be seen, as well as nuclear abnormalities as described in Figure III-79. Contrast this tissue fragment with those depicted in Figures III-72 and III-74. Papanicolaou stain, × 1000.

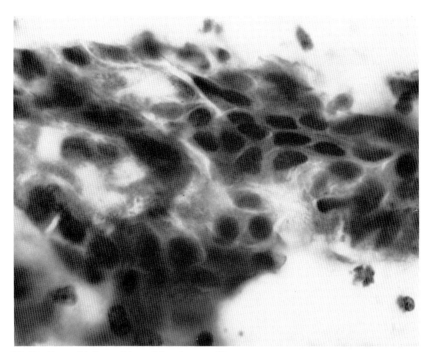

Fig. III-81. Bronchial washing. Membrane filter preparation. Squamous metaplasia with dysplasia. This was thought to be a severe dysplastic alteration arising in a metaplastic epithelium. In this fragment there is marked distortion of the architecture, with a pronounced increase in nuclear chromasia and individual nuclear cytoplasmic ratios. Papanicolaou stain, × 1000.

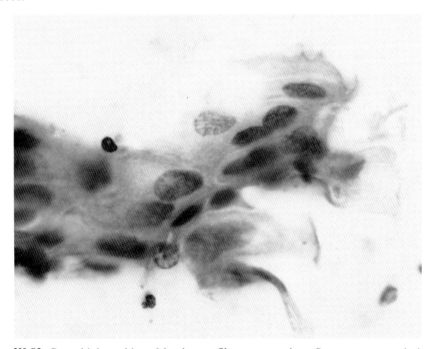

Fig. III-82. Bronchial washing. Membrane filter preparation. Squamous metaplasia with dysplasia. Papanicolaou stain, × 1000.

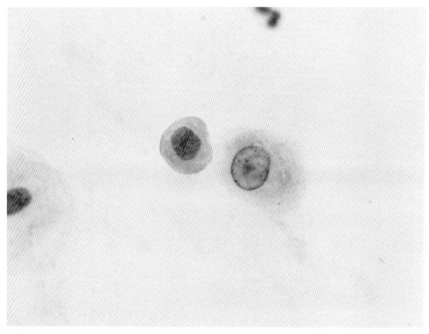

Fig. III-83. Fresh sputum. Squamous metaplasia with dysplasia. Papanicolaou stain, × 1000.

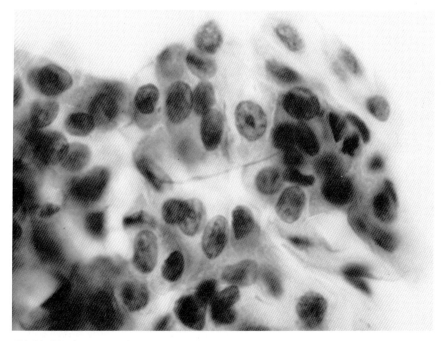

Fig. III-84. Fresh sputum. Squamous metaplasia with dysplasia. Papanicolaou stain, × 1000.

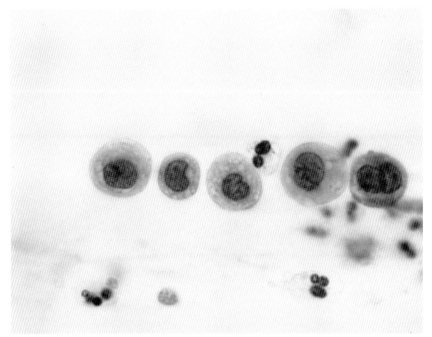

Fig. III-85. Fresh sputum. Squamous metaplasia with dysplasia. Papanicolaou stain, × 1000.

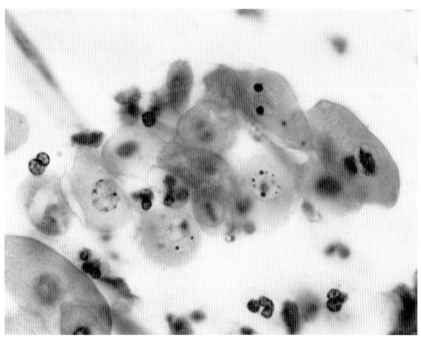

Fig. III-86. Fresh sputum. Squamous metaplasia with dysplasia and marked karyopyknosis and karyorrhexis. This is another field of the specimen in Figure III-85. A combination of the cells as noted in Figure III-85 and the necrotic changes as noted in Figure III-86 could lead to an erroneous diagnosis of squamous-cell carcinoma. The material in Figure III-86 is densely hyaline-like and orangeophilic. There is a marked resemblance to squamous ghosts from keratinizing squamous-cell carcinoma. Papanicolaou stain, × 1000.

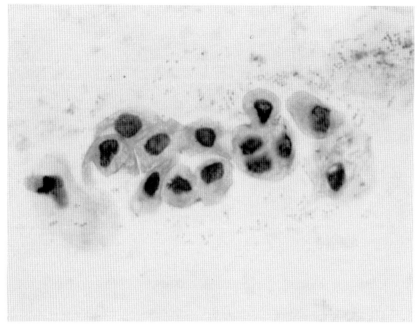

Fig. III-87. Fresh sputum. Squamous metaplasia. Papanicolaou stain, × 1000.

standing chronic irritation of the tracheobronchial tree, particularly by cigarette smoking; they have been seen to antedate the appearance of bronchogenic carcinoma; they have accompanied the cells of bronchogenic cancers of all types. Figures III-80 through III-85 depict examples of cellular changes which were interpreted as showing squamous metaplasia with focal dysplasia.

Cellular Alterations Resulting From Irradiation and Chemotherapy for Cancer

Profound alterations in epithelial cells from both the upper and the lower respiratory tract in response to irradiation therapy and anticancer chemotherapy have been described and reported from the earliest days in the development of respiratory cytopathology.[10] In the laboratories of the authors, multiple examples have been seen over the years. These cells may be so violently atypical and so easily mistaken for neoplastic cells that this special section is devoted to their description and illustration. The patients in whom we have most frequently observed cellular changes resulting from radiotherapy have been those who had previously received irradiation to the thorax for established bronchogenic carcinoma. Cytologic sampling of respiratory material had been continued to monitor these patients for tumor recurrence. Figures III-88 through III-92 depict the bizarre cells present in three different patients. The cells shown in Figures III-91 and III-92 were present in a bronchial brush specimen from a 49-year-old woman in whom a bronchiolo-alveolar cell carcinoma had been diagnosed seven months previously. Immediately following diagnosis the patient had received two 10-day courses of radiation, separated by an interval of one month, to the supramediastinal area. This history and the cellular changes have been fairly

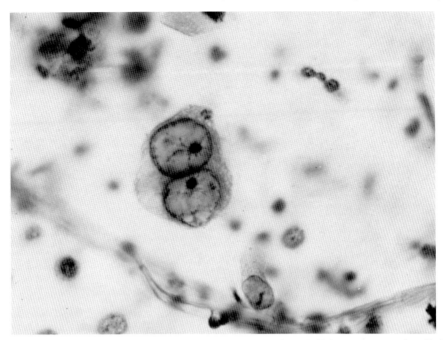

Fig. III-88. Bronchial washing. Membrane filter preparation. Cellular response to irradiation. Papanicolaou stain, × 1000.

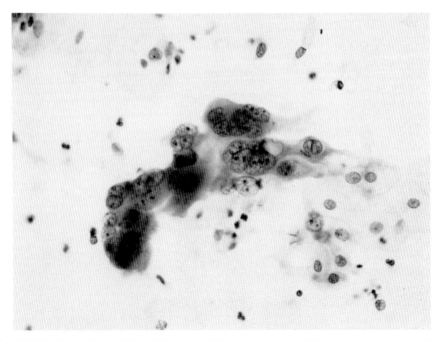

Fig. III-89. Bronchial washing. Membrane filter preparation. Cellular response to irradiation. Papanicolaou stain, × 400.

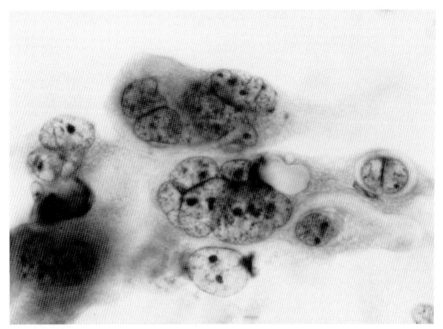

Fig. III-90. Bronchial washing. Membrane filter preparation. Cellular response to irradiation. Higher magnification of the cellular group shown in Figure III-89. The major characteristics of cellular response to irradiation are shown here; they are: cytomegaly with enlargement of nuclei and increase in cytoplasmic mass; multinucleation with the appearance of prominent and multiple nucleoli; and vacuolization of the cytoplasm. These are altered bronchial columnar cells. Papanicolaou stain, × 1000.

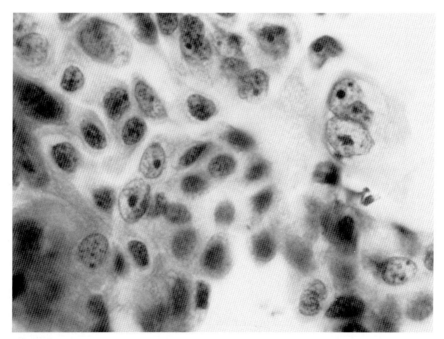

Fig. III-91. Bronchial brushing. Membrane filter preparation. Cellular response to irradiation. Papanicolaou stain, × 1000.

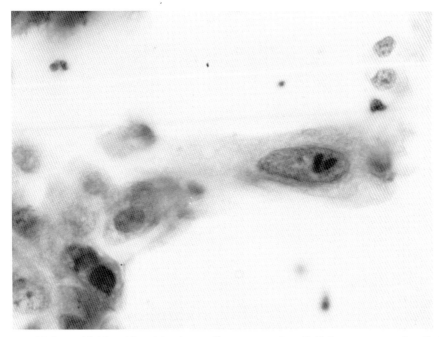

Fig. III-92. Bronchial brushing. Membrane filter preparation. Cellular response to irradiation. Papanicolaou stain, × 1000.

characteristic for all of our patients. Both squamous cells and columnar cells may be affected. These changes are characterized by cytomegaly, involving both cytoplasmic and nuclear enlargement, multinucleation, the emergence of prominent, sometimes huge, nucleoli, and cytoplasmic vacuolization.

The ever-increasing number of available anticancer drugs and their widespread utilization in cancer therapy have resulted in a greater frequency with which altered cells resulting from their use are found in specimens of respiratory material. Although cellular changes are seen most frequently following chemotherapy with busulfan, they have been described in association with most of the anticancer drugs, the most recent of which is bleomycin.[44] Specimens from three different patients exhibiting these bizarre cells are illustrated in Figures III-93 through III-100. The cells shown in Figures III-93 through III-96 were seen in bronchial brushing specimens from a 73-year-old woman who had had chronic myelogenous leukemia treated with busulfan intermittently for 11 years. A recent chest x-ray had shown a pulmonary infiltrate. It is of interest that at the time of the initial interpretation of the cytologic material, the history as given was not known. The cells were thought to be malignant. They were large, sparse, and present in sputum and in bronchial brushing specimens from several areas of the lung. They were characterized by hyperchromasia and macronucleoli. In occasional cells a degenerated ciliated border was present (Figure III-95). The authors have become convinced that a major key to correct recognition of these cells lies in the tendency of many of them to be roughly rectangular in shape (Figs. III-94 and III-96). Although these cells may arise from bronchial, bronchiolar, and alveolar epithelial cells, those in this case are most likely of bronchial columnar cell origin.

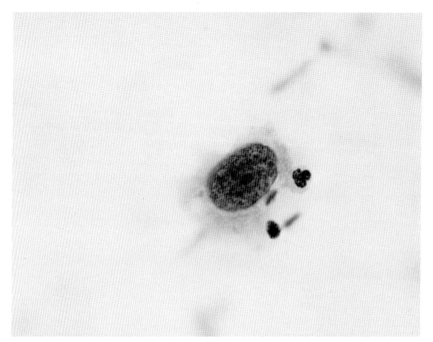

Fig. III-93. Bronchial brushing. Membrane filter preparation. Cellular response to anticancer chemotherapy. This is an extremely large cell with a very abnormal nucleus. There are multiple tiny invaginations of the nuclear membrane. The chromatin pattern is coarsely granular and quite hyperchromatic. There is one centrally placed macronucleolus. The cytoplasm is finely granular and is in a state of disintegration. Papanicolaou stain, × 1000.

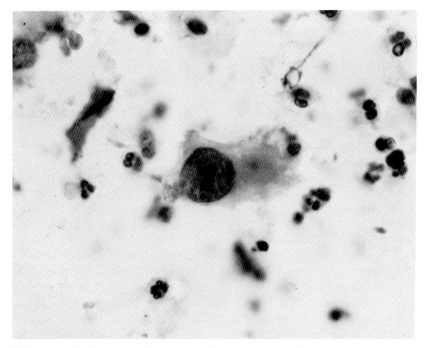

Fig. III-94. Bronchial brushing. Membrane filter preparation. Cellular response to anticancer chemotherapy. Another field of the specimen photographed in Figure III-93. Noteworthy is the rectangular shape of this very large cell. Papanicolaou stain, × 1000.

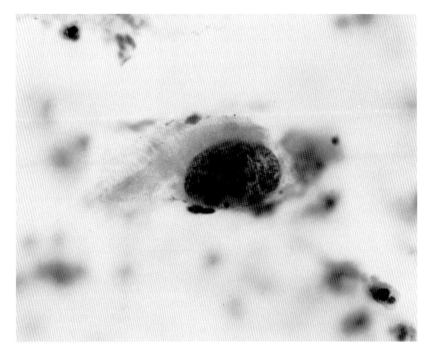

Fig. III-95. Bronchial brushing. Membrane filter preparation. Cellular response to anticancer chemotherapy. Another field from the specimen depicted in Figures III-93 and III-94. The nucleus is extremely abnormal, with a very densely hyperchromatic and granular pattern of chromatin. A distinct ciliated border is present, permitting a definite categorization of this cell as of bronchial origin. Papanicolaou stain, × 1000.

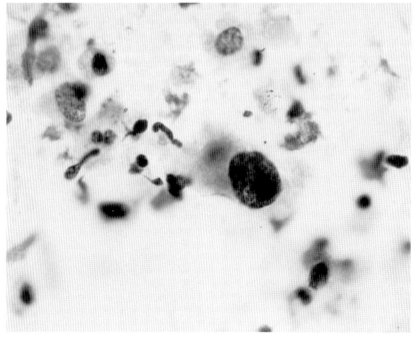

Fig. III-96. Bronchial brushing. Membrane filter preparation. Cellular response to anticancer chemotherapy. Another field from the specimen depicted in Figures III-93–95. Papanicolaou stain, × 1000.

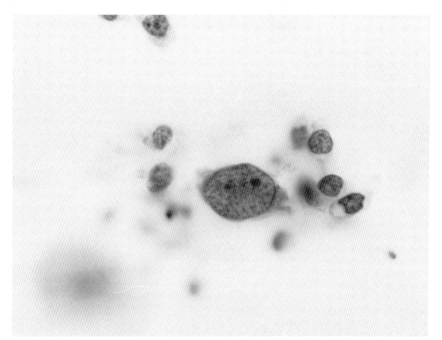

Fig. III-97. Bronchial washing. Membrane filter preparation. Cellular response to anticancer chemotherapy. Papanicolaou stain, × 1000.

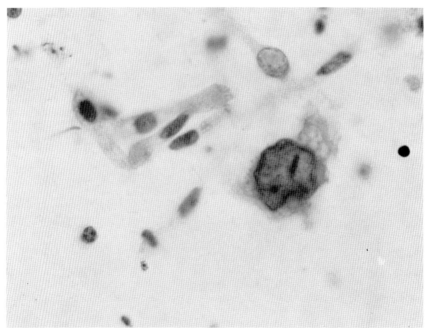

Fig. III-98. Bronchial washing. Membrane filter preparation. Cellular response to chemotherapy. Another field from the specimen shown in Figure III-97. Noteworthy is the huge size of the abnormal cell in comparison with normal bronchial ciliated columnar epithelial cells shown in the same photograph. Papanicolaou stain, × 1000.

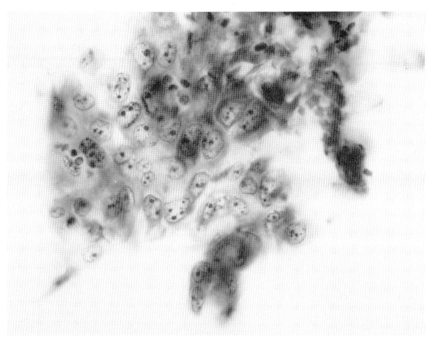

Fig. III-99. Bronchial washing. Membrane filter preparation. Cellular response to anticancer chemotherapy. Papanicolaou stain, × 400.

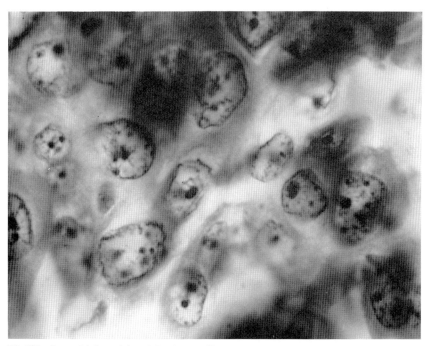

Fig. III-100. Bronchial washing. Membrane filter preparation. Cellular response to anticancer chemotherapy. Higher magnification of the tissue fragment shown in Figure III-99. At autopsy, the origin of this fragment was established to lie in altered bronchial columnar epithelium. Papanicolaou stain, × 1000.

The cells in Figures III-97 and III-98 were seen in the bronchial brushing specimen from a 52-year-old man with a history of chronic myelogenous leukemia. He had been treated with several different chemotherapeutic drugs and with corticosteroids, and was thought to be in blast crises at the time of bronchoscopy. The cellular changes present in the specimen were reported to be consistent with cellular reaction to chemotherapy.

Figures III-99 and III-100 show at two magnifications a markedly abnormal tissue fragment, which was present in a bronchial washing specimen from an 18-year-old man with lymphomatoid granulomatosis that had been treated with massive chemotherapy and corticosteroids. An autopsy revealed profound and widespread alterations in the bronchial epithelium.

CELLS OF NONEPITHELIAL ORIGIN

The Pulmonary Alveolar Macrophage

The pulmonary alveolar macrophage and its biologic significance to the cytopathologist have recently been summarized by Frost and his associates.[11] Like alterations of the bronchial mucosa, abnormalities of these cells are becoming of increasing interest in studies of noxious environmental inhalants.[10, 45-47] The presence of these cells is mandatory in establishing the satisfactory nature of a sputum specimen. These macrophages are recognized by the eccentric position of the nucleus, barely touching the cytoplasmic membrane (Figs. III-101 and III-102). There may be abundant

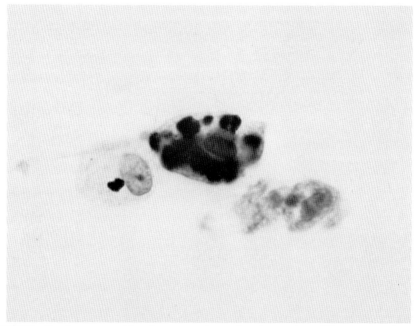

Fig. III-101. Fresh sputum. Alveolar macrophages bearing carbon. Three macrophages are depicted, all of which contain large aggregates of black pigment, presumably carbon. The cytoplasm is finely vacuolated and the nuclei occupy an eccentric position. Indentation of the nucleus on the side facing toward the center of the cell is commonly observed. One or more stainable nucleoli may be seen. The chromatin pattern is uniform and finely granular. Papanicolaou stain, × 1000.

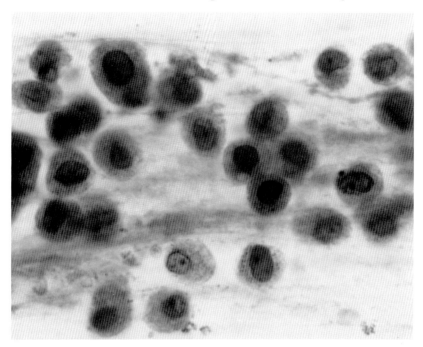

Fig. III-102. Fresh sputum. Alveolar macrophages bearing carbon. Papanicolaou stain, ×
1000.

foamy cytoplasm and phagocytized material, usually carbon. On occasion the nuclei
may assume a bean shape and show one or more nucleoli. Other cytoplasmic
inclusions have been noted in the presence of environmental pollutants. Binucleated
and multinucleated giant macrophages are not infrequently encountered (Figs. III-
103 and III-104). Nuclei may vary in number from four or five to several hundred.
Typical foreign-body type or Langhans-type giant cells may be seen. These cells may
be identified in association with chronic lung disease of many varieties, including
sarcoidosis, tuberculosis, and other granulomatous diseases, but they are not indica-
tive of any of these and may be seen in the sputum in the absence of clinical disease.
A case in point may be pulmonary sarcoidosis. In a recent study from the Johns
Hopkins Hospital, sputum cytology from 16 patients with histologically confirmed
sarcoidosis was reviewed. Cytologic findings thought to be diagnostically significant
included multinucleated giant cells, some containing Schaumann and asteroid bodies,
epithelioid cells, and lymphoid cells.[48] Giant cells and epithelioid cells in relation to
pulmonary tuberculosis are discussed further in Chapter IV.

Large vacuoles containing fat have been reported in macrophages in the presence
of lipoid pneumonia.[49, 50] In Figures III-105 through III-108 are shown fat-laden
macrophages from a patient with lipoid pneumonia, most likely related to aspiration
of nose drops containing oil. On first observation these cells indeed present quite an
ominous appearance. However careful study reveals that the large size of the cells
and the multiple vacuoles are responsible for their startling character. The nuclei,
although multiple in some cells, are small and bland. Correct identification of these
features should lead to avoidance of a diagnosis of malignancy, particularly adenocarcinoma or liposarcoma.

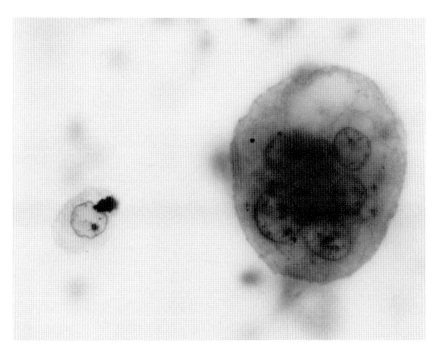

Fig. III-103. Bronchial brush. Membrane filter preparation. Multinucleated alveolar macrophage. Papanicolaou stain, × 1000.

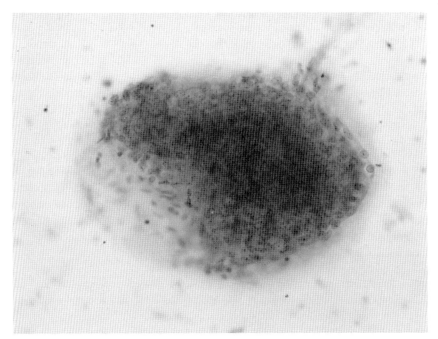

Fig. III-104. Bronchial brush. Membrane filter preparation. Multinucleated alveolar macrophage. This cell illustrates rather dramatically the enormous number of nuclei that may be observed. There is some suggestion of a Langhans giant cell because of the tendency of the nuclei to concentrate more toward one border of the cell. The authors in their laboratories have been unable to make any correlation between such cells and the presence of any granulomatous disease. Papanicolaou stain, × 250.

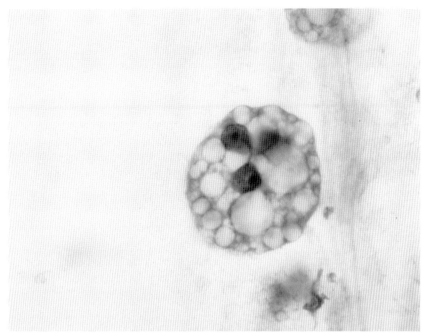

Fig. III-105. Fresh sputum. Alveolar macrophage, containing lipid from a patient with lipoid pneumonia. Of diagnostic significance is the presence of multiple, thick-walled vacuoles present in a cell which otherwise bears the characteristics of an alveolar macrophage. Papanicolaou stain, × 1000.

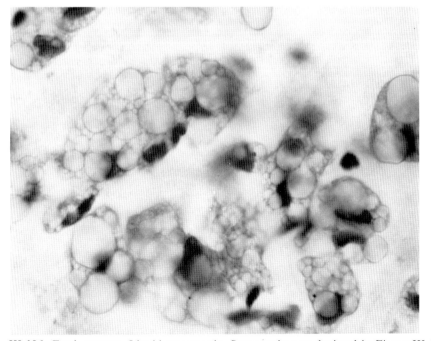

Fig. III-106. Fresh sputum. Lipoid pneumonia. Same patient as depicted in Figure III-105. Papanicolaou stain, × 1000.

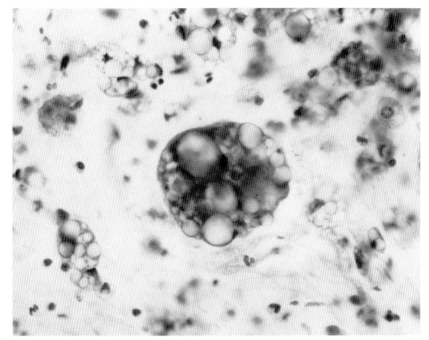

Fig. III-107. Fresh sputum. Lipoid pneumonia. Same patient as depicted in Figures III-105 and III-106. Papanicolaou stain, × 400.

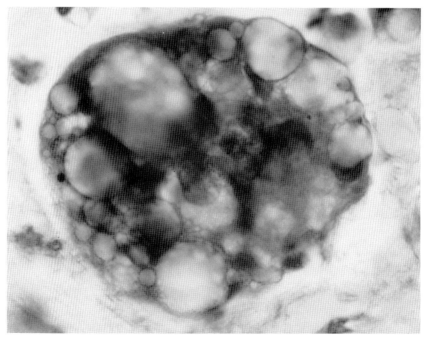

Fig. III-108. Fresh sputum. Lipoid pneumonia. Same patient as depicted in Figures III-105–107. Papanicolaou stain, × 1000.

Pulmonary alveolar macrophages bearing hemosiderin may be seen in specimens of sputum and bronchial material in any situation in which there has been an escape of erythrocytes into the alveolar spaces. Its presence has greater significance than fresh blood, because it denotes chronicity and rules out recent instrumentation as a possible cause of bleeding. As an aid in the diagnosis of one unusual clinical entity, idiopathic pulmonary hemosiderosis, the presence of macrophages bearing hemosiderin is of great importance. One of the authors (Johnston) has seen several patients with this disease. In one child the macrophages were recovered from swallowed sputum obtained by gastric aspiration. A typical cell is shown in Figure III-109. The hemosiderin pigment is present in large, rounded aggregates, which are usually refractile and orange or rust colored. Lowering the substage condenser will increase this refractility. When diagnostic recognition is critical, the presence of hemosiderin should be confirmed by use of a special stain for iron.

Lymphoid Cells

Other cells that originate from circulating blood and may be seen in bronchopulmonary material include lymphocytes of varying degrees of maturity.[51] They may be associated with a chronic inflammatory process or with the rupture of a lymphoid follicle in the wall of the bronchus. The lymphocytes often stream out in the strands

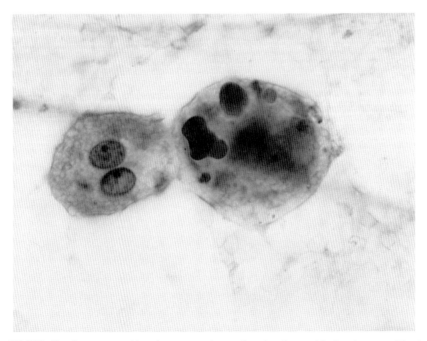

Fig. III-109. Fresh sputum. Alveolar macrophages bearing hemosiderin pigment. The hemosiderin is characterized by occurrence in large, rounded masses, which may coalesce. The staining may vary from a bright orange to a rusty brown color. Identification may be enhanced by the racking down of the substage condenser, which increases the refractility of these masses. With such a manipulation other masses such as carbon and melanin pigment should become less refractile and grow dark. Specific identification of hemosiderin, however, should be based only on an appropriate special stain. Papanicolaou stain, × 1000.

of mucus, mimicking the exfoliation pattern of small-cell anaplastic carcinoma. The lymphocytes may be distinguished from the small-cell undifferentiated neoplasm by their failure to mold to one another, lack of intercellular recognition, and lack of malignant criteria. Often when a follicle is ruptured, one will also find phagocytic reticulum cells and capillaries present, and a diagnosis of chronic follicular bronchitis may be made. Lymphocytes are shown in Figures III-110 and III-111.

Eosinophils

Although most frequently associated with asthmatic bronchitis, the presence of eosinophils in pulmonary material may be noted with any disease of the lower respiratory tract in which there is a component of allergy. These diseases may include, in addition to asthma, eosinophilic pneumonia of multiple etiologies, Wegener's granulomatosis, and sarcoidal angiitis.[52] In material stained by the Papanicolaou technique, eosinophils may be readily identified, or they may be difficult to recognize. This inconstant pattern is due to the lack of specific affinity of the counterstains in the Papanicolaou method for the cytoplasmic granules of the eosinophils. Eosinophils in sputum stained by the Papanicolaou method are shown in Figure III-112. The cytoplasmic granules are only faintly visible. Correct recognition is aided by the characteristic bilobed nucleus.

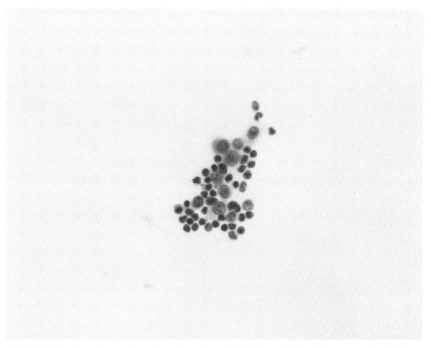

Fig. III-110. Fresh sputum. Lymphocytes mixed with bronchial columnar cells. At times the differential diagnostic interpretation of small cells in respiratory specimens can become critical. Lymphocytes occur as single cells or as cells that delicately touch one another with slight indentation. Intense molding should not be present. The nuclear chromatin should be uniform and may exhibit marked hyperchromasia, although stainable nucleoli are very rare or absent. Papanicolaou stain, × 400.

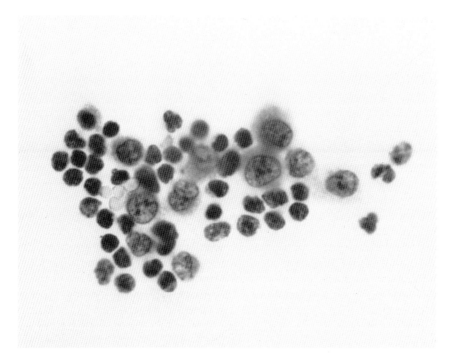

Fig. III-111. Fresh sputum. Lymphocytes mixed with bronchial columnar cells. This is a higher magnification of Figure III-110. Papanicolaou stain, × 1000.

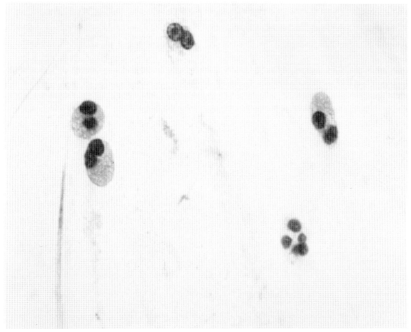

Fig. III-112. Fresh sputum. Eosinophils. Identification of eosinophils in respiratory material may be extremely easy or very difficult because of the indifference of the Papanicolaou staining procedure to the cytoplasmic granules. Careful examination of this figure will reveal faint outlines of cytoplasmic granules, but no distinct staining. The bilobed nuclei are characteristic. Specific identification of eosinophils, when critical, should be based upon utilization of a Romanowsky stain. Papanicolaou stain, × 1000.

ENDOGENOUS AND EXOGENOUS NONCELLULAR AND NONLIVING COMPONENTS

A number of different nonliving structures may be present in cytologic specimens from the lower respiratory tract. The presence of some may indicate specific pulmonary problems; others may serve only to confuse and produce incorrect diagnoses. Some may be derived from the patient; others may contaminate the specimen after it has been taken from the patient.

Mucus

Curschmann's spirals are casts of small bronchi formed from mucus showing varying degrees of inspissation. Although they are most frequently associated with asthmatic bronchitis, these spirals may be seen in the presence of any condition in which the plugging of small bronchi with mucus is encouraged. Thus their presence in a pulmonary specimen denotes bronchial obstruction. Such blockage of the small bronchi could prevent the appearance of neoplastic cells in the cytologic specimen. Treatment of the patient with bronchodilators will usually reduce this obstruction and permit the passage of neoplastic cells. Figures III-113 through III-115 depict these spirals.

Occasionally small inspissated masses of mucus will round up and adhere to one another so as to suggest nuclei. In extreme situations (Fig. III-116) the nuclear hyperchromasia and molding of small-cell undifferentiated carcinoma may be mimicked. Absence of any intranuclear detail or of any cytoplasmic rim should be the features leading to correct interpretation.

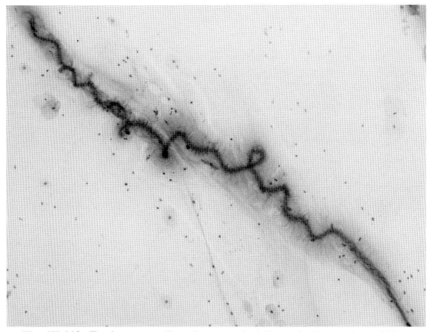

Fig. III-113. Fresh sputum. Curschmann spiral. Papanicolaou stain, × 100.

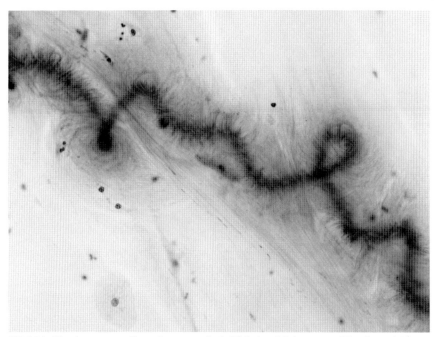

Fig. III-114. Fresh sputum. Curschmann spiral. This is a higher magnification of Figure III-113. Papanicolaou stain, × 250.

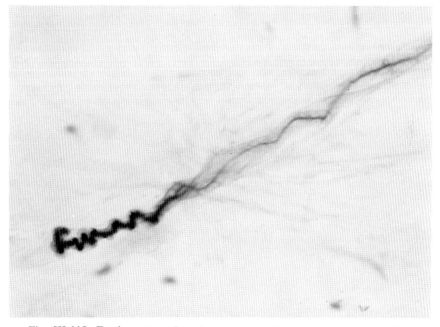

Fig. III-115. Fresh sputum. Curschmann spiral. Papanicolaou stain, × 250.

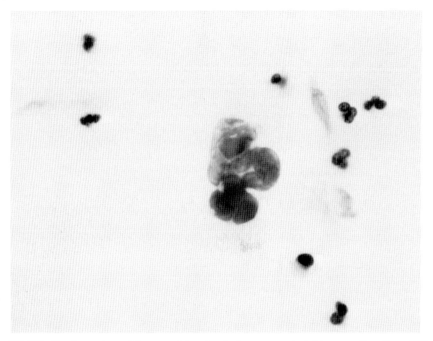

Fig. III-116. Fresh sputum. Masses of inspissated mucus mimicking small-cell undifferentiated carcinoma. Papanicolaou stain, × 1000.

Ferruginous Bodies

Ferruginous bodies have been noted in pulmonary tissues and cytologic specimens for many years. They were formerly called *asbestos bodies,* reflecting the belief that they were all formed as a reaction to inhaled fibers of asbestos. Now it is known that a number of different inhaled mineral fibers result in identical structures.

These bodies are composed of various substances, including iron, which are encrusted upon a thin, needle-like mineral fiber (Figs. III-117 through III-120). They are most commonly mistaken for fungi by the beginner. Increasing attention is being focused upon their relationship to bronchogenic carcinoma. In the authors' institutions a report of the presence of ferruginous bodies in pulmonary specimens is a signal to the clinician for closer surveillance of the patient for possible early lung cancer.

Corpora Amylacea and Calcospherites

Several varieties of rounded bodies may be formed in the respiratory passages and alveoli and subsequently appear in specimens of sputum or bronchial material. Through the years these bodies have been given different names, but those most frequently used now are corpora amylacea, psammoma bodies, and calcospherites. The terms psammoma body and calcospherite are generally considered to be synonyms. Both corpora amylacea and calcospherites appear as rounded homogenous masses varying in size between 30 and 200 microns. Both present concentric rings and radial striations (Figs. III-121 through III-124), but chemically they are different.

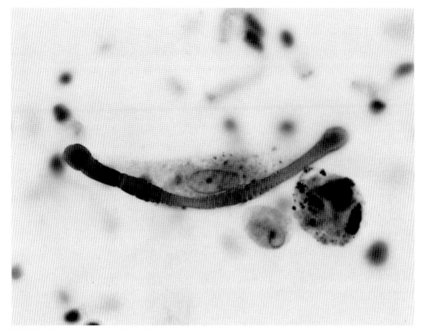

Fig. III-117. Bronchial washing. Membrane filter preparation. Ferruginous body with reacting macrophages. Papanicolaou stain, × 1000.

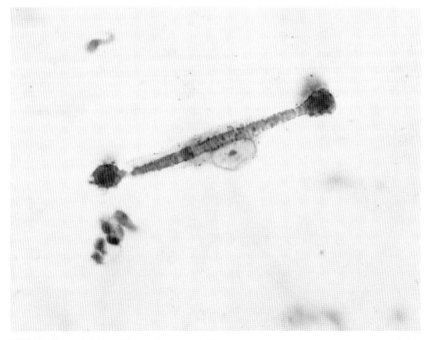

Fig. III-118. Bronchial washing. Membrane filter preparation. Ferruginous body with reacting macrophage. Papanicolaou stain, × 1000.

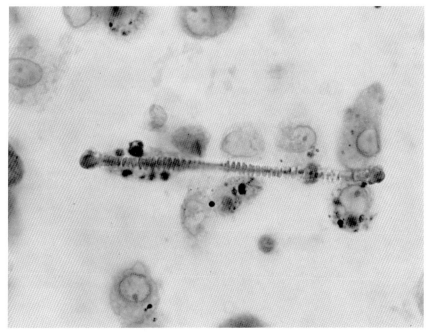

Fig. III-119. Bronchial washing. Membrane filter preparation. Ferruginous body with reacting macrophages. Papanicolaou stain, × 1000.

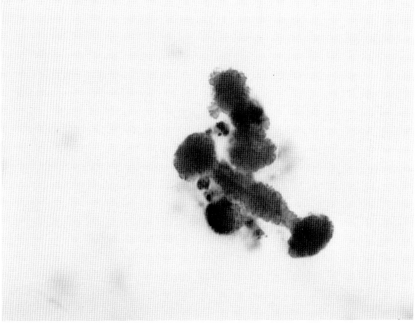

Fig. III-120. Fresh sputum. Ferruginous body. Note the particularly intense encrustation of mineral deposits. Papanicolaou stain, × 1000.

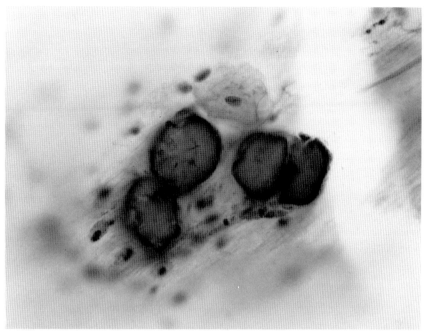

Fig. III-121. Fresh sputum. Corpora amylacea. Papanicolaou stain, × 400.

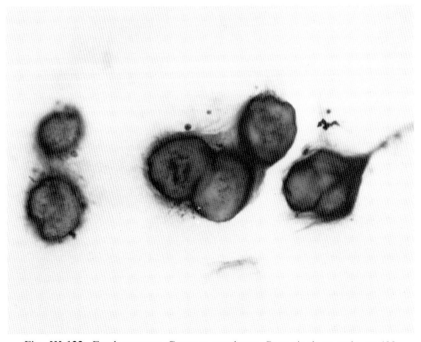

Fig. III-122. Fresh sputum. Corpora amylacea. Papanicolaou stain, × 400.

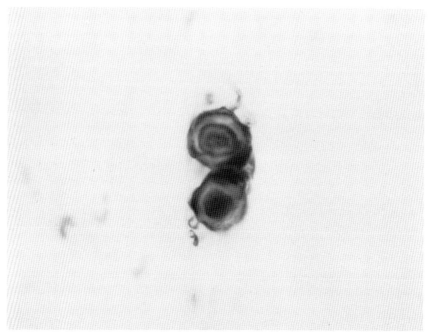

Fig. III-123. Fresh sputum. Calcospherites. Papanicolaou stain, × 1000.

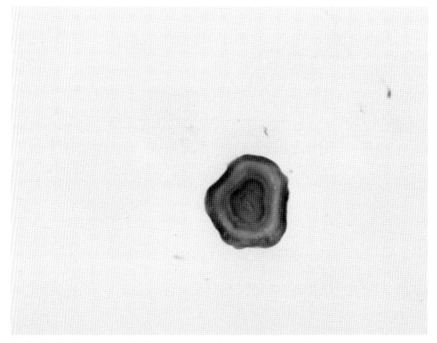

Fig. III-124. Fresh sputum. Calcospherites. These are also referred to as psammoma bodies and are characterized by a very distinct laminated pattern and marked basophilia secondary to the staining of deposited calcium. Papanicolaou stain, × 1000.

Corpora amylacea for the most part are composed of glycoproteins. They do not calcify. In contrast, calcospherites are calcified and contain phosphates, iron, magnesium, and sudanophilic material.[52] Corpora amylacea are seen particularly under circumstances of heart failure, pulmonary infarction, and chronic bronchitis. Calcospherites have been associated with the rare disease pulmonary microlithiasis and with malignant neoplasms capable of producing these bodies.

Contaminants

It is beyond the scope of this monograph to discuss exhaustively all of the structures capable of contaminating bronchopulmonary material. Those introduced in this section are included because of the possibility of their misinterpretation, resulting in erroneous diagnoses. Figures III-125 through III-130 depict examples of plant cells, originating in food or other sources, which have appeared in cytologic specimens. The most common source of these cells is the spaces between the teeth of the patient. To the beginner such cells can appear quite menacing. Indeed the authors have seen students, residents, and inexperienced cytotechnologists misdiagnose them as neoplastic cells from various types of cancers. The recognition of cellulose cell walls (Figs. III-125 through III-128) and cytoplasmic granulations (Figs. III-128 through III-130) is a clue to the vegetable origin of the cells.

Figure III-131 shows a granule of starch from glove powder embedded in a mass of cellular debris in a specimen of sputum. These granules may bear a superficial

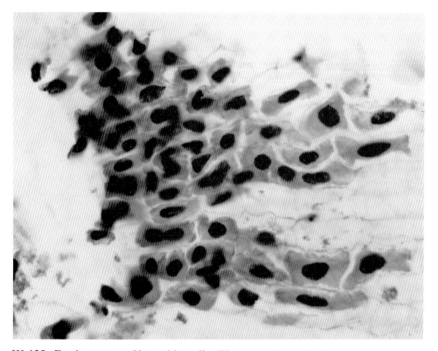

Fig. III-125. Fresh sputum. Vegetable cells. The cytoplasm of these cells stains intensely orange with the Papanicolaou stain. The beginner could easily mistake this pattern for that of keratinizing squamous-cell carcinoma. Papanicolaou stain, × 400.

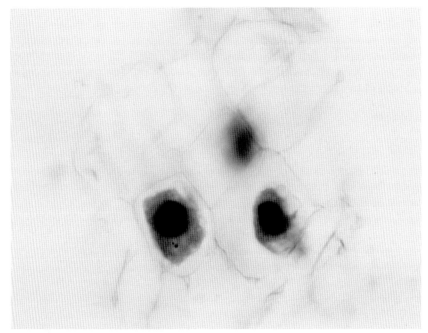

Fig. III-126. Fresh sputum. Vegetable cells. The cellulose membranes are well preserved here. Papanicolaou stain, × 1000.

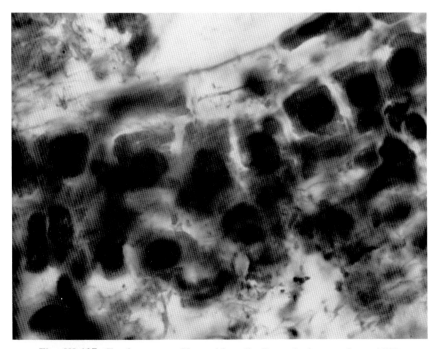

Fig. III-127. Fresh sputum. Vegetable cells. Papanicolaou stain, × 1000.

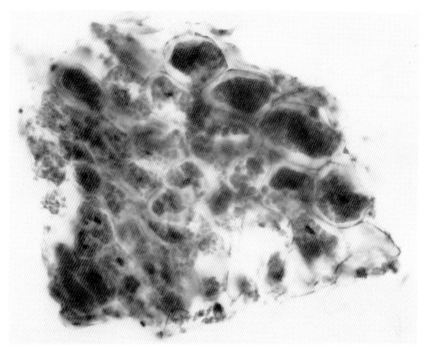

Fig. III-128. Fresh sputum. Vegetable cells. Note the large cytoplasmic granulations. Papanicolaou stain, × 1000.

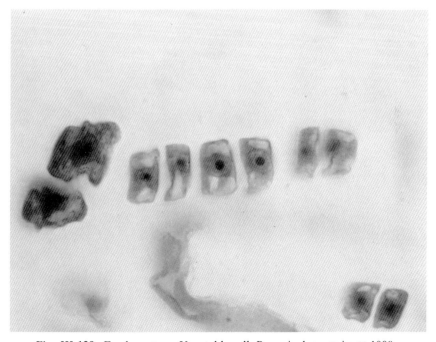

Fig. III-129. Fresh sputum. Vegetable cell. Papanicolaou stain, × 1000.

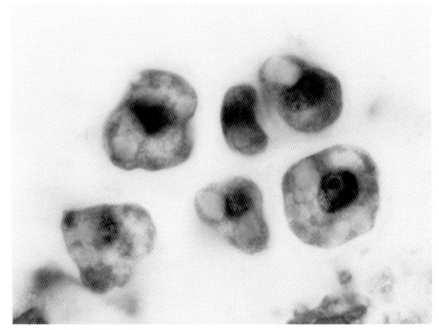

Fig. III-130. Fresh sputum. Vegetable cells. Papanicolaou stain, × 1000.

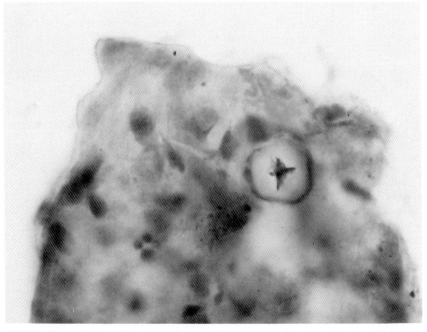

Fig. III-131. Bronchial washing. Membrane filter preparation. Starch granule mimicking cryptococcus. Papanicolaou stain, × 1000.

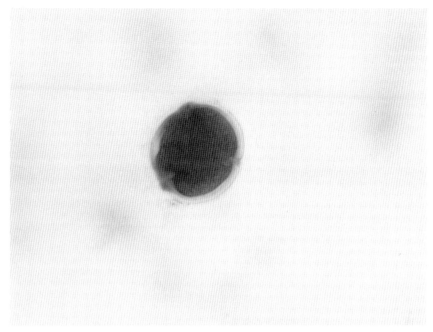

Fig. III-132. Fresh sputum. Pollen granule. Papanicolaou stain, × 1000.

resemblance to *Cryptococcus neoformans.* See Chapter IV for further discussion of this differential diagnostic problem.

A microspore of pollen is shown in Figure III-132. The thick refractile wall with indentations or projections is characteristic. This structure should not be confused with the various yeast-like mycotic agents described in Chapter IV.

REFERENCES

1. Farber, S. M., Wood, D. A., Pharr, S. L., and Pierson, B.: Significant cytologic findings in nonmalignant pulmonary disease. *Dis Chest 31:* 1–13, 1957.
2. Woolner, L. B. and McDonald, J. R.: Bronchogenic carcinoma: Diagnosis by microscopic examination of sputum and bronchial secretions: Preliminary report. *Proc Staff Meetings Mayo Clin 22:* 369–381, 1947.
3. Woolner, L. B. and McDonald, J. R.: Diagnosis of carcinoma of the lung: The value of cytologic study of sputum and bronchial secretions. *JAMA 139:* 497–502, 1949.
4. Woolner, L. B. and McDonald, J. R.: Cytologic diagnosis of bronchogenic carcinoma. *Am J Clin Pathol 19:* 765–769, 1949.
5. Woolner, L. B. and McDonald, J. R.: Carcinoma cells in sputum and bronchial secretions: A study of 150 consecutive cases in which results were positive. *Surg Gynecol Obstet 88:* 273–290, 1949.
6. Woolner, L. B. and McDonald, J. R.: Cytologic diagnosis of bronchogenic carcinoma. *Dis Chest 17:* 1–10, 1950.
7. Woolner, L. B. and McDonald, J. R.: Cytology of sputum and bronchial secretions: Studies on 588 patients with miscellaneous pulmonary lesions. *Ann Intern Med 33:* 1164–1174, 1950.
8. McDonald, J. R.: Exfoliative cytology in genitourinary and pulmonary diseases. *Am J Clin Pathol 24:* 684–687, 1954.
9. McDonald, J. R.: Pulmonary cytology. *Am J Surg 89:* 462–464, 1955.
10. Koss, L. G.: *Diagnostic Cytology and Its Histopathologic Basis.* J. B. Lippincott, Philadelphia, 1968.
11. Frost, J. K., Gupta, P. K., Erozan, Y. S., Carter, D., Hollander, D. H., Levin, M. L., and Ball, W. C., Jr: Pulmonary cytologic alterations in toxic environmental inhalation. *Hum Pathol 4:* 521–536, 1973.

12. Kinsella, D. L.: Bronchial cell atypias: A report of a preliminary study correlating cytology with histology. *Cancer 12:* 463–472, 1959.
13. Nasiell, M.: The general appearance of the bronchial epithelium in bronchial carcinoma: A histopathological study with some cytological viewpoints. *Acta Cytol 7:* 97–106, 1963.
14. Kierszenbaum, A. L.: Bronchial metaplasia: Observations on its histology and cytology. *Acta Cytol 9:* 365–371, 1965.
15. Nasiell, M.: Metaplasia and atypical metaplasia in the bronchial epithelium: A histopathologic and cytopathologic study. *Acta Cytol 10:* 421–427, 1966.
16. Nasiell, M.: Abnormal columnar cell findings in bronchial epithelium: A cytologic and histologic study of lung cancer and noncancer cases. *Acta Cytol 11:* 397–402, 1967.
17. Plamenac, P. and Nikulin, A.: Atypia of the bronchial epithelium in wind instrument players and in singers: A cytopathologic study. *Acta Cytol 13:* 274–278, 1969.
18. Saccomanno, G., Saunders, R. P., Klein, M. G., Archer, V. E., and Brennan, L.: Cytology of the lung in reference to irritant, individual sensitivity, and healing. *Acta Cytol 14:* 377–381, 1970.
19. Plamenac, P., Nikulin, A., and Pikula, B.: Cytology of the respiratory tract in former smokers. *Acta Cytol 16:* 256–260, 1972.
20. Plamenac, P., Nikulin, A., and Pikula, B.: Cytologic changes of the respiratory tract in young adults as a consequence of high levels of air pollution exposure. *Acta Cytol 17:* 241–244, 1973.
21. Plamenac, P., Nikulin, A., and Pikula, B.: Cytologic changes of the respiratory epithelium in iron foundry workers. *Acta Cytol 18:* 34–40, 1974.
22. Hoch-Ligeti, C. and Eller, L. L.: Significance of multinucleated epithelial cells in bronchial washings. *Acta Cytol 7:* 258–261, 1963.
23. Chalon, J., Katz, J. S., Ramannthon, S., Ambirunga, M., and Orkin, L. R.: Tracheobronchial epithelial multinucleation in malignant diseases. *Science 183:* 525–526, 1974.
24. Kawecka, M.: Cytological evaluation of sputum in patients with bronchiectasis and the possibilities of erroneous diagnosis of cancer. *Acta Unio Int Contra Cancrum 15:* 469, 1959.
25. Naylor, B. and Railey, C.: A pitfall in the cytodiagnosis of sputum of asthmatics. *J Clin Pathol 17:* 84–89, 1964.
26. Sanerkin, N. G. and Evans, D. M. D.: The sputum in bronchial asthma: Pathognomonic patterns. *J Pathol Bacteriol 89:* 535–541, 1965.
27. Williams, J. W.: Alveolar metaplasia: Its relationship to pulmonary fibrosis in industry and development of lung cancer. *Br J Cancer 11:* 30–42, 1957.
28. Berkheiser, S. W.: Bronchiolar proliferation and metaplasia associated with bronchiectasis, pulmonary infarcts, and anthracosis. *Cancer 12:* 499–508, 1959.
29. Berkheiser, S. W.: Bronchiolar proliferation and metaplasia associated with thromboembolism. A pathological and experimental study. *Cancer 16:* 205–211, 1963.
30. Kern, W. H.: Cytology of hyperplastic and neoplastic lesions of terminal bronchioles and alveoli. *Acta Cytol 9:* 372–379, 1965.
31. Masin, F. and Masin, M.: Frequencies of alveolar cells in concentrated sputum specimens related to cytologic classes. *Acta Cytol 10:* 362–367, 1966.
32. Cooney, W., Dzuira, B., Harper, R., and Nash, G.: The cytology of sputum from thermally injured patients. *Acta Cytol 16:* 433–437, 1972.
33. Valentine, E. H.: Squamous metaplasia of the bronchus. A study of metaplastic changes occurring in the epithelium of the major bronchi in cancerous and non-cancerous cases. *Cancer 10:* 272–279, 1957.
34. Auerbach, O., Gere, J. B., Forman, J. B., Petrick, T. G., Smolin, H. J., Muehsam, G. E., Kassouny, D. Y., and Stout, A. P.: Changes in the bronchial epithelium in relation to smoking and cancer of the lung: A report of progress. *N Engl J Med 256:* 97–104, 1957.
35. Carroll, R.: Changes in the bronchial epithelium in primary lung cancer. *Br J Cancer 15:* 215–219, 1961.
36. Ford, D. K., Fidler, H. K., and Lock, D. R.: Dysplastic lesions of the bronchial tree. *Cancer 14:* 1226–1234, 1961.
37. Auerbach, O., Stout, A. P., Hammond, E. C., and Garfinkel, L.: Changes in bronchial epithelium in relation to cigarette smoking and in relation to lung cancer. *N Engl J Med 265:* 253–267, 1961.
38. Koprowska, I., An, S. H., Corsey, D., Dracopoulos, I., and Vaskelis, P. S.: Cytologic patterns of developing bronchogenic carcinoma. *Acta Cytol 9:* 424–430, 1965.
39. Saccomanno, G., Saunders, R. P., Archer, V. E., Auerbach, O., Kuschner, M., and Beckler, P. A.: Cancer of the lung: The cytology of sputum prior to the development of carcinoma. *Acta Cytol 9:* 413–423, 1965.
40. Fullmer, C. D., Short, J. G., Allen, A., and Walker, K.: Proposed classification for bronchial epithelial cell abnormalities in the category of dyskaryosis. *Acta Cytol 13:* 459–471, 1969.
41. Saccomanno, G., Archer, V. E., Auerbach, O., Saunders, R. P., and Brennan, L. M.: Development of carcinoma of the lung as reflected in exfoliated cells. *Cancer 33:* 256–270, 1974.

42. Neweoehner, D. E., Kleinerman, J., and Rice, D. B.: Pathologic changes in peripheral airways of young cigarette smokers. *N Engl J Med 291:* 755–758, 1974.
43. Naib, Z. M.: Pitfalls in the cytologic diagnosis of oat cell carcinoma of the lung. *Acta Cytol 8:* 34–38, 1964.
44. Bedrossian, C. W. M. and Corey, B. J.: Abnormal sputum cytopathology during chemotherapy with bleomycin (abstract). *Acta Cytol 20:* 586, 1976.
45. An, S. H. and Koprowska, I.: Primary cytologic diagnosis of asbestosis associated with bronchogenic carcinoma: Case report and review of literature. *Acta Cytol 6:* 391–398, 1962.
46. Roque, A. L. and Pickren, J. W.: Enzymatic changes in fluorescent alveolar macrophages of the lungs of cigarette smokers. *Acta Cytol 12:* 420–429, 1968.
47. Walker, K. R. and Fullmer, C. D.: Observations of eosinophilic extracytoplasmic processes in pulmonary macrophages: Progress report. *Acta Cytol 15:* 363–364, 1971.
48. Aisner, S. C., Gupta, P. K., and Frost, J. K.: Sputum cytology in pulmonary sarcoidosis. *Acta Cytol 21:* 394–398, 1977.
49. Losner, S., Volk, B. W., Slade, W. R., Nathanson, L., and Jacobi, M.: Diagnosis of lipid pneumonia by examination of sputum. *Am J Clin Pathol 20:* 539–545, 1950.
50. Masin, F. and Masin, M.: Sputum concentration technic with assessment of sudanophilic lipids. *Acta Cytol 10:* 134–137, 1966.
51. Tassoni, E. M.: Pools of lymphocytes: Significance in pulmonary secretions. *Acta Cytol 7:* 168–173, 1963.
52. Spencer, H.: *Pathology of The Lung.* WB Saunders, Philadelphia, 1977.

IV   Infectious Diseases

The awareness that cytologic technique can play a significant role in the primary diagnosis of infectious diseases has been steadily increasing.[1-7] The diseases in which cytology has been of greatest use are those in which the responsible microorganism has a morphology specific enough to be detected by light microscopic methods or to produce specific cellular changes.

MYCOBACTERIAL INFECTIONS

Cytopathology generally has had very little to offer in the category of bacterial infection. A number of studies have been devoted to the exfoliations associated with pulmonary tuberculosis.[8-12] Indeed, on occasion in patients with tuberculosis, one may observe epithelioid cells and Langhan giant cells. One of the authors (Johnston) recently encountered two such cases in his own laboratory. In one case, shown in Figures IV-1 and IV-2, collections of macrophages in a necrotic background were interpreted as possibly representing epithelioid cells and caseous necrotic material. Some of the nuclear shapes in these cells have been characterized by Nasiell as "carrot-shaped". A tissue fragment observed in the bronchial brushings from a second patient (Figs. IV-3 and IV-4) was interpreted as suggestive of a noncaseating tuberculoid granuloma. Indeed, both of these patients were found to have active pulmonary tuberculosis. Unfortunately such cytologic findings appear to be rare. Likewise searches for acid-fast bacilli in cytologic material from these patients have proved disappointing. On only one previous occasion were we able to restain a sputum smear for acid-fast bacilli and detect in several macrophages the presence of acid-fast organisms. These occurred in a child with *Mycobacterium intracellulari* (Battey bacillus). Attempts by us to apply this observation to other patients with atypical mycobacterial disease have been unsuccessful.

MYCOSES

The role of cytology in the diagnosis of mycotic infections has become increasingly valuable. In these diseases, the etiologic agent is visible and in many cases has a

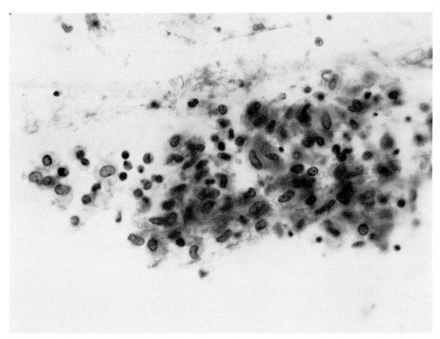

Fig. IV-1. Fresh sputum. Epithelioid cells and caseous necrosis from active pulmonary tuberculosis. This diagnosis was made cytologically without prior evidence of the presence of tuberculosis. Papanicolaou stain, × 400.

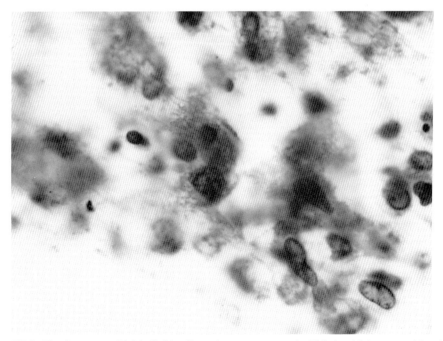

Fig. IV-2. Fresh sputum. Epithelioid cells and caseous necrosis. This is a higher magnification of the cells depicted in Figure IV-1. Papanicolaou stain, × 1000.

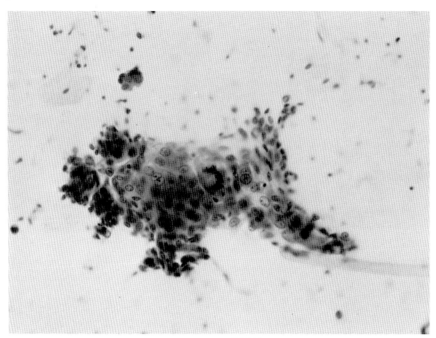

Fig. IV-3. Bronchial brushing. Membrane filter preparation. Noncaseating tuberculoid granuloma from patient with active pulmonary tuberculosis. A Langhans giant cell is present in the center of the granuloma. Papanicolaou stain, × 250.

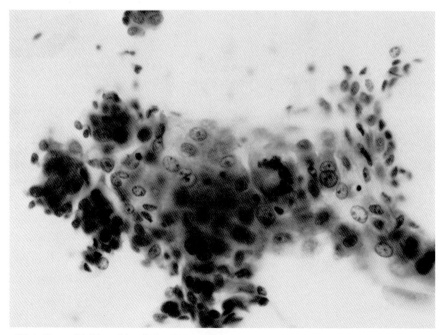

Fig. IV-4. Bronchial brushing. Membrane filter preparation. Noncaseating tuberculoid granuloma from patient with active pulmonary tuberculosis. Higher magnification of Figure IV-3. Papanicolaou stain, × 400.

morphology on which a specific diagnosis may be based. The detection of these organisms in a stained cytologic specimen may be the first clue to the nature of the patient's problems. The accuracy of observation is dependent upon the ability of the cytotechnologist and the pathologist to appreciate the various forms that these organisms may assume. Specific morphology and refractility of cell walls are the most important criteria for detection. Staining characteristics are too variable to be of help. Although special stains for fungi are useful, with several exceptions they are not necessary. Indeed it has been established that these organisms are easily seen on a Papanicolaou stain. Mucicarmine or alcian blue help to visualize the capsule of cryptococcus, but the organism can be identified on the basis of other features. Plane of focus may be another critical factor on which detection depends. Some of the fungi are so thick that they may be blurred or even absent in the plane of focus at which human cells are being studied on the smear or membrane filter.

In the following pages are presented discussions of those organisms which the authors have encountered in their laboratories and which are most common. The listing is not intended to be all-inclusive or exhaustive; rather, it is an attempt to set forth some general guidelines for diagnosis.

Blastomyces dermatitidis

North American blastomycosis is an infectious disease caused by the fungus *Blastomyces dermatitidis* and characterized by the formation of granulomatous and suppurative lesions in any part of the body. Of protean possible manifestations, the disease is frequently first seen as a lung mass but may not be recognized until after the production of skin or subcutaneous lesions. The symptoms, signs, and x-ray changes may closely resemble the appearance and progression of bronchogenic carcinoma. Symptoms may consist of cough, dyspnea, chest pain, low-grade fever, weight loss, and weakness. The sputum may become purulent or blood-streaked. In well-developed cases the x-rays show unilateral, dense, irregular shadows, which frequently are indistinguishable from those produced by an infiltrating malignant neoplasm.

This organism has been frequently encountered in our laboratories. The following case histories are typical of the clinical setting in which pulmonary material is submitted for cytologic evaluation.

Case 1. An 85-year-old woman, two years prior to her admission, had had an accute episode of pneumonic infiltration, which had never completedly cleared and which by x-ray had been slowly increasing for the past year. This had been accompanied by low-grade fever, malaise, chronic cough, and weight loss. As the patient had refused any surgical intervention, her physician had elected to follow her through repeated sputum cytologic examination and x-rays. It was his impression that she had a malignant tumor of the lung. One week before her admission a sputum submitted for routine cytologic interpretation revealed many thick-walled, single-budding yeast forms highly character-istic of *Blastomyces dermatitidis.* Shortly thereafter, this observation was confirmed by positive culture identification. The patient was then admitted for definitive therapy. One month later the yeast could no longer be found in the sputum, and the patient was markedly improved.

Case 2. A 52-year-old man was admitted for evaluation of a left hilar mass and left-sided pneumonia. The patient had been in good health, with the exception of a history of chronic alcoholism, until six months prior to admission, at which time he began to develop malaise, fatiguability, and occasional night sweats. Anorexia and a 20-pound weight loss occurred. Two weeks prior to admission, the patient became increasingly ill with nausea and vomiting, cough, fever, chills, and increase in sputum production. On admission chest film showed a large mass in the left lung. Initially, the patient was thought to have bronchogenic carcinoma with a secondary pneumonia. Five days later a sputum cytologic smear revealed budding yeasts characteristic for *Blastomyces dermatitidis.* Blastomyces was also cultured from the sputum. Amphotericin B therapy was instituted with subsequent gradual, but marked, clinical improvement both in the patient's clinical state and in the chest film.

Case 3. A 52-year-old woman, approximately two months prior to her admission, had been seen in the emergency room of another hospital with the complaint of diffuse right anterolateral chest pain. A chest x-ray showed two small infiltrates in the right lung. A complete series of diagnostic studies failed to reveal the etiology of the pulmonary infiltrate, and she was referred to Duke University Medical Center for further evaluation. Three early-morning sputum specimens and one bronchial washing were reported as negative for malignant cells or infectious agent. At thoracotomy biopsy of the right upper lobe of the lung and the chest wall was carried out. Microscopic examination of this tissue revealed an intense granulomatous reaction with the formation of small tubercles and focal accumulations of neutrophils. Many characteristic budding yeast forms of *Blastomyces dermatitidis* were seen. A review of the four cytopathologic specimens revealed the presence of blastomyces in one of the four.

Case 4. A 62-year-old man was transferred from another hospital where he had been treated with steroids for diabetes control. Chest x-ray at that time revealed fluffy white infiltrates bilaterally. He was begun on penicillin and steroids. Cytologic examination of sputum revealed *Blastomyces dermatitidis.* He expired within a few days and autopsy revaled systemic blastomycosis.

Case 5. A 58-year-old man had had polymyositis for 5–6 years. He had received steroid therapy and was admitted to the hospital with cough, fever, and a small infiltrate in the right lung. He was also having exacerbation of his polymyositis. Sputum was taken in the first few days of admission. He followed a relentless course with cough, fever, and marked increase in pulmonary infiltrate. Bronchial brush and tracheal aspirates were taken the day before his death. *Blastomyces dermatitidis* was present in all specimens.

In cytopathologic materials fixed in 95% alcohol and stained by the Papanicolaou technique, *Blastomyces dermatitidis* appears as single or budding spherical cells, 8–15 microns in diameter, with thick, refractile walls (Figs. IV-5 through IV-10). The thickness of the walls along with some tendency for the cell mass to retract away from these walls may impart to most forms a "double-contoured" appearance. No hyphae are seen. The most important criterion for morphologic confirmation of blastomyces is the nature of the budding. Single budding is characteristic. The bud has a tendency to remain in close apposition to the mother cell, so that a flattening of the two apposed surfaces occurs. Staining is highly variable and of little use as an aid in differential diagnosis. The wall is quite refractile and may stain a pale blue-green. The cytoplasm stains variably. In some cells scattered granules of varying staining qualities may be seen embedded in an otherwise nonstaining cytoplasmic mass. Ultrastructural examination has revealed that these small masses are multiple nuclei. In other cells the entire cytoplasmic mass may shrink within the cell wall and stain basophilicly. This shrinkage produces a halo between the cytoplasmic mass and the cell wall, a useful characteristic aiding in identification. The unwary may well believe these yeast cells are of human origin, the cell wall being mistaken for cytoplasm and the cytoplasmic mass for a nucleus.[13] In a more recent study, Craig,

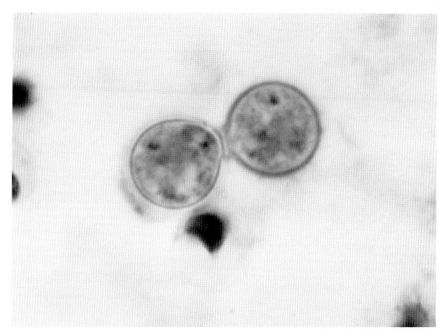

Fig. IV-5. Fresh sputum. *Blastomyces dermatitidis.* Major diagnostic features of the organism are depicted here. The walls are thick and highly refractile, with a double contoured appearance. The budding is single and broad-based. Papanicolaou stain, × 1000.

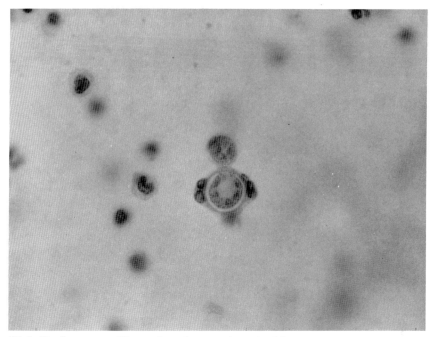

Fig. IV-6. Fresh sputum. *Blastomyces dermatitidis.* A budding yeast form is shown with the halo artefact produced by shrinkage of the cell body away from the cell wall. Papanicolaou stain, × 1000.

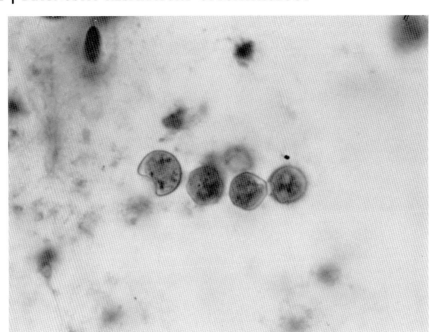

Fig. IV-7. Fresh sputum. *Blastomyces dermatitidis.* Three organisms are shown, one of which is budding. In the organism to the extreme left there has been an inward collapse of the cell wall. Papanicolaou stain, × 1000.

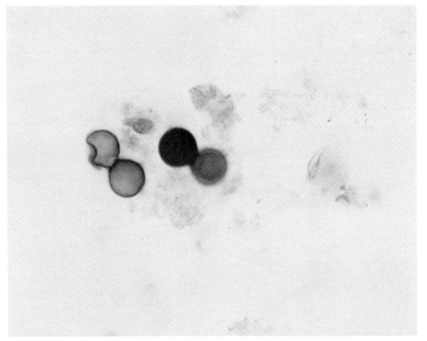

Fig. IV-8. Fresh sputum. *Blastomyces dermatitidis.* Two budding organisms are present. Papanicolaou stain, × 1000.

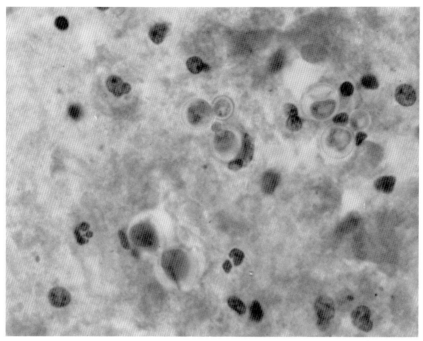

Fig. IV-9. Fresh sputum. *Blastomyces dermatitidis.* Multiple budding and nonbudding yeast forms are present, embedded in a background of necrotic debris and mixed inflammatory exudate. Papanicolaou stain, × 1000.

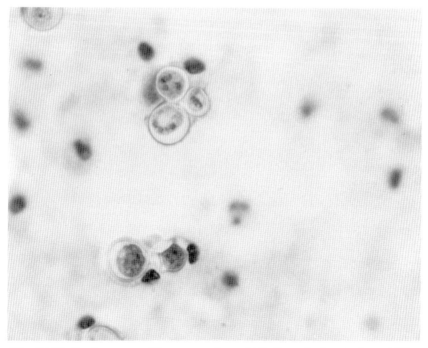

Fig. IV-10. Fresh sputum. *Blastomyces dermatitidis.* Papanicolaou stain, × 1000.

Nair, and Bluestein[14] have emphasized the usefulness of electron-microscopical observations on specimens of sputum as an aid in the differential diagnostic problems surrounding blastomyces.

The other fungi from which blastomyces must be differentiated are *Cryptococcus neoformans, Coccidioides immitis,* and *Histoplasma capsulatum. Cryptococcus neoformans* also buds singly, but the single bud pinches off, leaving a markedly attenuated isthmus of attachment to the mother cell and thus assuming a teardrop shape. *Coccidioides immitis* may be seen in cytologic material as a round, double-walled cell measuring from 10 to over 100 microns in diameter, but it does not bud and may be seen to contain yellowish-brown endospores.[2] *Histoplasma capsulatum* is an extremely small budding yeast, 1 to 5 microns in diameter, and is rarely seen outside of macrophages.

Inflammatory reaction induced by infection with *Blastomyces dermatitidis* may vary from that of the production of classic tuberculoid granulomata to the formation of microabscesses with a predominance of the neutrophil. The cytologic examination of bronchopulmonary material from such cases reflects this spectrum of inflammatory reactivity. Many polymorphonuclear leukocytes are invariably present. Occasionally, multinucleated giant cells may also be seen. The budding yeast of blastomyces may appear within these giant cells; in such cases the diagnosis of the disease is confirmed (Figs. IV-11 through IV-20); however, more frequently giant cells without apparent organisms are observed. There is nothing specific about the cellular picture to suggest that infection with blastomyces has occurred.

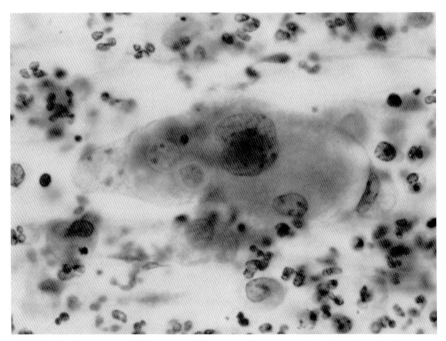

Fig. IV-11. Fresh sputum. *Blastomyces dermatitidis.* A large macrophage bearing several nuclei is shown. A phagocytized organism is present in the left-hand portion of the cell. Papanicolaou stain, × 1000.

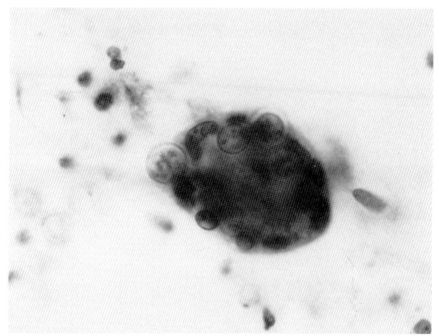

Fig. IV-12. Fresh sputum. *Blastomyces dermatitidis.* A large, multinucleated macrophage bearing several organisms is present. Papanicolaou stain, × 1000.

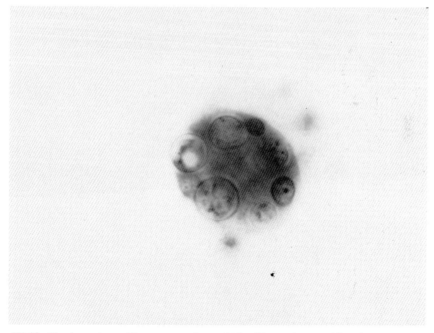

Fig. IV-13. Fresh sputum. *Blastomyces dermatitidis.* Shown is a multinucleated macrophage bearing multiple organisms. Papanicolaou stain, × 1000.

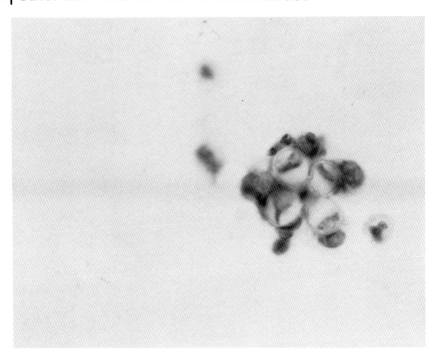

Fig. IV-14. Fresh sputum. *Blastomyces dermatitidis.* A cluster of organisms surrounded by neutrophils and macrophages is present. Papanicolaou stain, × 1000.

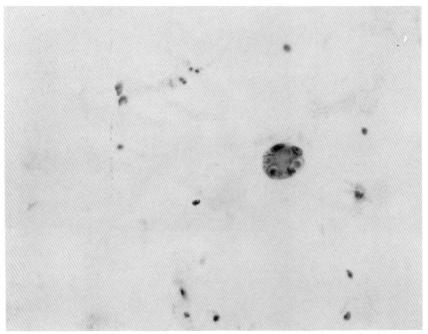

Fig. IV-15. Fresh sputum. *Blastomyces dermatitidis.* Several organisms phagocytized by a multinucleated macrophage are present. Papanicolaou stain, × 400.

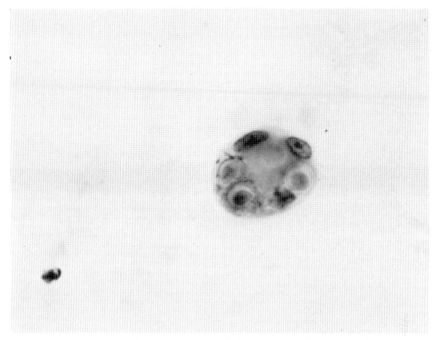

Fig. IV-16. Fresh sputum. *Blastomyces dermatitidis.* This is a higher magnification of Figure IV-15. Papanicolaou stain, × 1000.

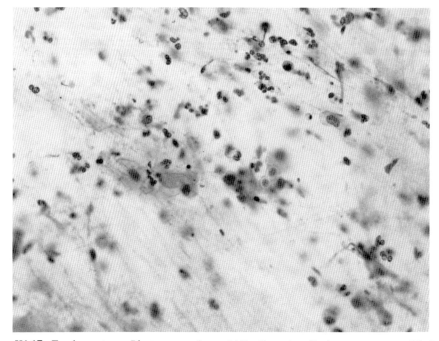

Fig. IV-17. Fresh sputum. *Blastomyces dermatitidis.* Occasionally large amounts of inflammatory exudate with necrosis may make visualization of organisms more difficult. One may be seen just to the right of center. Papanicolaou stain, × 400.

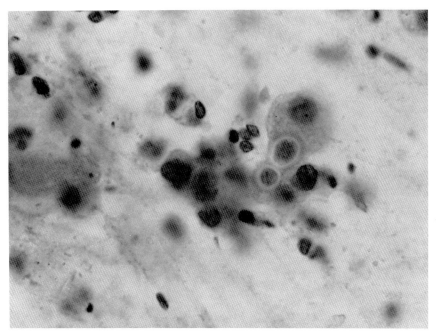

Fig. IV-18. Fresh sputum. *Blastomyces dermatitidis.* Shown at higher magnification is the organism depicted in Figure IV-17. Papanicolaou stain, × 1000.

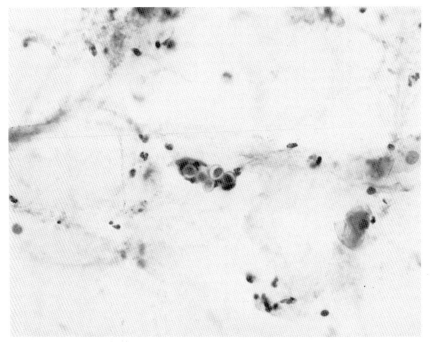

Fig. IV-19. Fresh sputum. *Blastomyces dermatitidis.* In the center of the field is a cluster of organisms surrounded by macrophages. Papanicolaou stain, × 400.

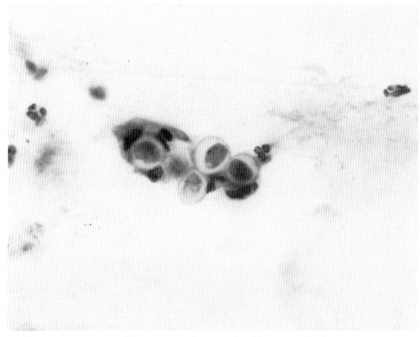

Fig. IV-20. Fresh sputum. *Blastomyces dermatitidis.* Shown at higher magnification is the cluster of organisms depicted in Figure IV-19. Papanicolaou stain, × 1000.

Cryptococcus neoformans

Although this organism is capable of producing primary disease in man, it is most frequently seen as a secondary invader in a patient whose immune responses have been compromised either by prior disease or by therapy. The following case history illustrates a frequently occurring clinical setting.

> A 56-year-old man five months prior to admission had received a cadaveric kidney transplantation. Following discharge he did well on high-dose steroids, but gradually developed increasing dyspnea and cough. X-rays revealed multiple pulmonary infiltrates. Sputum cytologies were obtained and revealed the presence of *Cryptococcus neoformans.*

Although the authors have most frequently observed these budding yeast forms in sputum and in bronchial material, we have seen them in specimens of urine and cerebrospinal fluid as well. Like blastomyces, single budding is characteristic of this yeast; but in contrast to that of blastomyces, the cryptococcus single bud pinches off, leaving a markedly attenuated isthmus of attachment to the mother cell and thus assuming a tear drop shape. Each cell is ovoid to spherical with a thickened wall and measures 5–20 microns in diameter, but there may be a much more marked variation in size of these organisms than is seen with blastomyces. Whereas the blastomyces are usually at the moderate to large end of the size spectrum, cryptococci are frequently very small. As a consequence they may be easily overlooked. Additionally, the cryptococci are much more variable in their internal morphology than the blastomyces. These variable morphologic presentations are shown in Figures IV-21 through IV-34 and are described in more detail in the legends. The organism is often

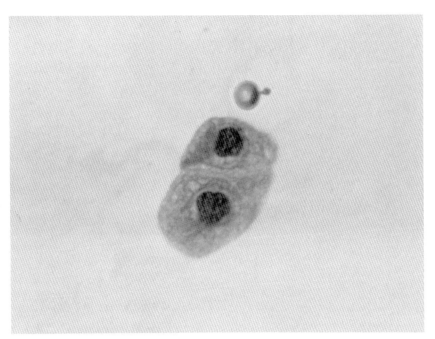

Fig. IV-21. Fresh sputum. *Cryptococcus neoformans.* The organism is present in the figure just above the two metaplastic cells. Papanicolaou stain, × 1000.

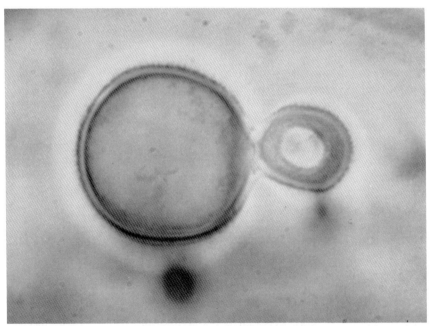

Fig. IV-22. Fresh sputum. *Cryptococcus neoformans.* In this figure a number of the major diagnostic features of this organism are depicted. It buds singly, with the buds forming a thin isthmus of attachment and assuming a tear-drop shape. Frequently, as shown in this figure, the interior of the organism has a rather vague and empty look, in marked contrast to *Blastomyces dermatitidis.* Compare this figure with Figure IV-21, taken at the same magnification, in order to fully appreciate the striking contrast in size these organisms can exhibit. Papanicolaou stain, × 1000.

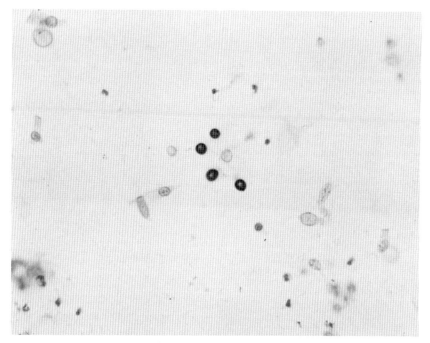

Fig. IV-23. Bronchial brushing. Membrane filter preparation. *Cryptococcus neoformans.* Four nonbudding yeast forms are present in the center of the field. Papanicolaou stain, × 400.

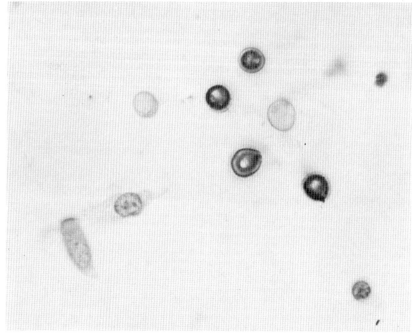

Fig. IV-24. Bronchial brush. Membrane filter preparation. *Cryptococcus neoformans.* This figure is a higher magnification of the organisms depicted in Figure IV-23. Note the presence of the rather amorphous, dark centers in these organisms. This characteristic is frequently observed. Although it is more pronounced on membrane filters, it can also be seen in smear preparations. Papanicolaou stain, × 1000.

111

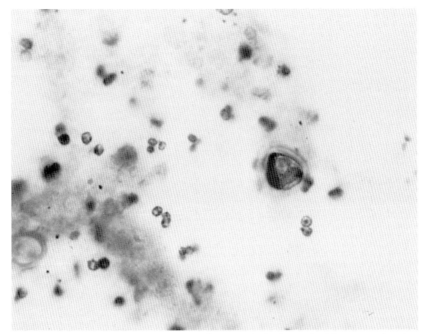

Fig. IV-25. Fresh sputum. *Cryptococcus neoformans.* A single nonbudding organism is present slightly to the right of the center and is surrounded by a mixed inflammatory exudate. Note the same dark, amorphous center as was depicted in Figures IV-23 and IV-24. Papanicolaou stain, × 1000.

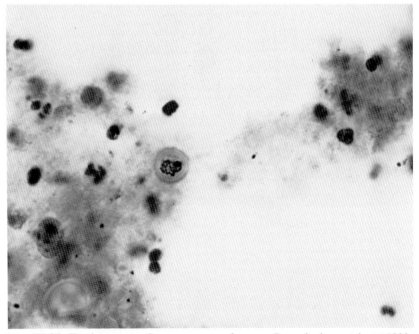

Fig. IV-26. Fresh sputum. *Cryptococcus neoformans.* Papanicolaou stain, × 1000.

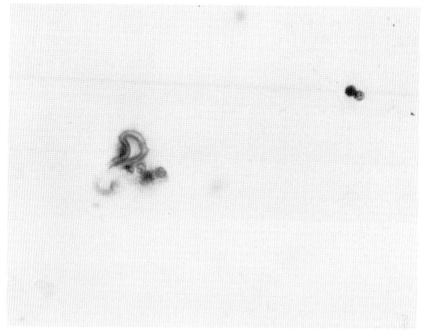

Fig. IV-27. Fresh sputum. *Cryptococcus neoformans.* To the left of center is a nonbudding form. The cell wall has collapsed inwardly. This is a frequently observed feature in these organisms. Papanicolaou stain, × 1000.

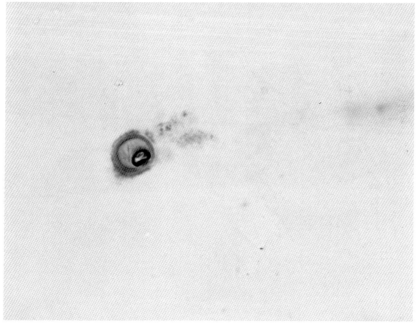

Fig. IV-28. Fresh sputum. *Cryptococcus neoformans.* Note the dark crystalloid artefact in the center of the yeast form. Papanicolaou stain, × 1000.

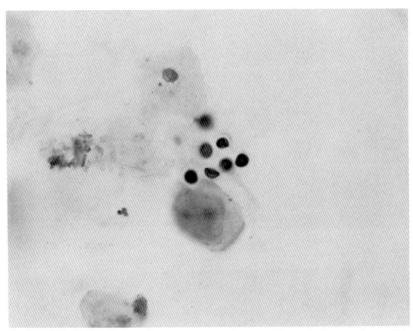

Fig. IV-29. Fresh sputum. *Cryptococcus neoformans.* Papanicolaou stain, × 400.

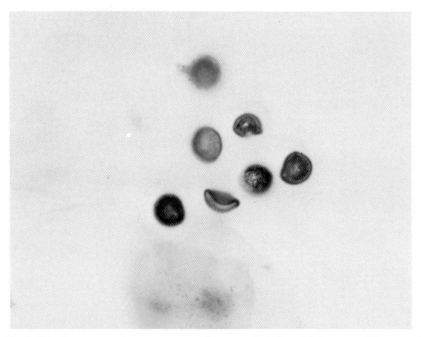

Fig. IV-30. Fresh sputum. *Cryptococcus neoformans.* This is a higher magnification of the organisms depicted in Figure IV-29. Multiple, collapsed yeast forms are present. Papanicolaou stain, × 1000.

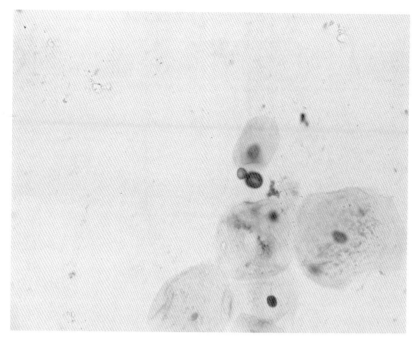

Fig. IV-31. Fresh sputum. *Cryptococcus neoformans.* A single budding yeast is present. Papanicolaou stain, × 400.

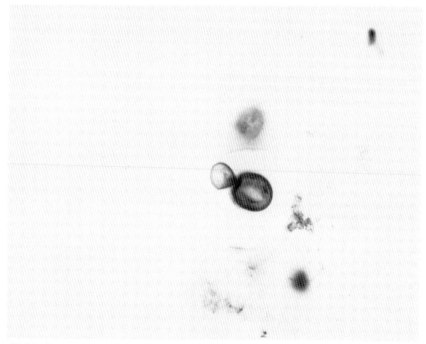

Fig. IV-32. Fresh sputum. *Cryptococcus neoformans.* This figure shows a higher magnification of the organism depicted in Figure IV-31. Although the cryptococcus is characterized by the ability to elaborate a thick mucopolysaccharide capsule, this may not always be present in Papanicolaou-stained smears. In this particular field a capsule can be observed as it indents the cytoplasm of the surrounding squamous cells. Papanicolaou stain, × 1000.

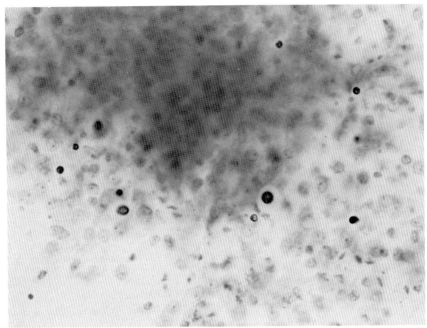

Fig. IV-33. Fresh sputum. *Cryptococcus neoformans.* Multiple organisms are present. The surrounding halos represent capsular material. Papanicolaou stain, × 1000.

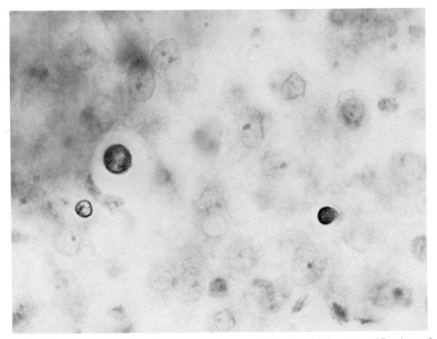

Fig. IV-34. Fresh sputum. *Cryptococcus neoformans.* This is a higher magnification of the field depicted in Figure IV-33. Papanicolaou stain, × 1000.

surrounded by a gelatinous capsule, which usually requires a special stain for visualization. However, on occasion, these capsules may stain with the Papanicolaou technique. Not infrequently this organism may be seen in sputum as well as elsewhere, with a virtual absence of inflammatory exudate.[3, 4, 6, 15-17]

Coccidioides immitis

> *Case History.* A 36-year-old woman was seen for evaluation of bilateral upper-lobe infiltrates noted on x-ray two months prior to admission. Her history was significant in that she had had episodes of gross hemoptysis for four years with no medical evaluation. Of further significance was that as a military dependent she had lived in a number of parts of the United States and had been in Hawaii and California. Cytologic examination of sputum revealed spherules of *Coccidioides immitis.*

In this case history the significant information is the presence of pulmonary infiltrates in a patient who has lived in geographical areas in which *Coccidioides immitis* is endemic. Because of the increasing nomadic habits of people in the United States, an infected patient may be seen in almost any hospital.

Spherules and endospores of coccidioides have been reported also in cytologic preparations of gastric washings and cerebrospinal fluid, as well as in sputum and bronchial washings. The spherules of *Coccidioides immitis* (Figs. IV-35 through IV-38) may present a particularly dramatic appearance because of their ability to assume sizes in excess of 100 microns in diameter (Fig. IV-35). The spherule may be empty or may contain endospores. The latter are round, nonbudding structures measuring

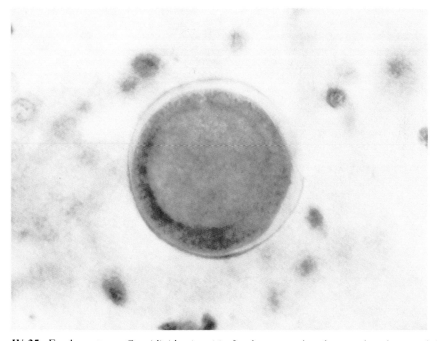

Fig. IV-35. Fresh sputum. *Coccidioides immitis.* In the center is a large spherule containing many endospores. Papanicolaou stain, × 1000.

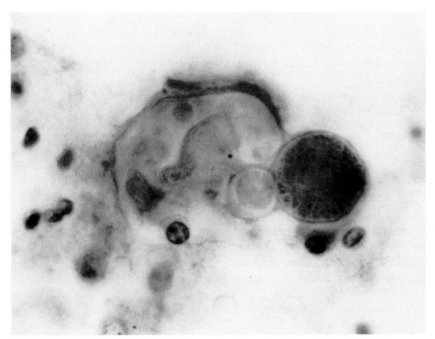

Fig. IV-36. Fresh sputum. *Coccidioides immitis.* Several spherules bearing endospores are present. The proximity of spherules suggesting budding is an artefact. Papanicolaou stain, × 1000.

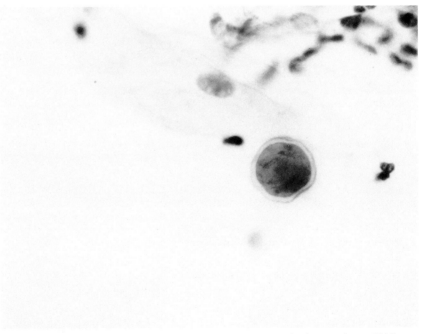

Fig. IV-37. Fresh sputum. *Coccidioides immitis.* Papanicolaou stain, × 1000.

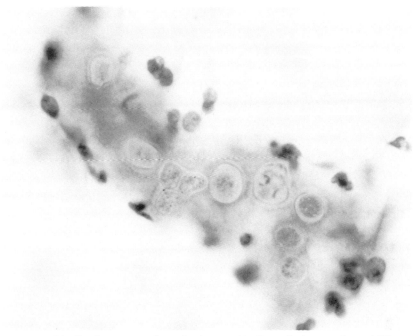

Fig. IV-38. Fresh sputum. *Coccidioides immitis.* Papanicolaou stain, × 1000.

2–5 microns in diameter (Fig. IV-36). The empty spherules may be confused with nonbudding forms of *Blastomyces dermatitidis.* Arthrospores may be encountered and must be differentiated from those of geotrichum species.[18]

The spherules of *Coccidioides immitis* may be encountered quite frequently in the sputum of infected patients and are particularly common in sputum of patients in those parts of the country in which the disease is endemic.

Histoplasma capsulatum

Figures IV-39 and IV-40 show the organisms of *Histoplasma capsulatum* stained with methenamine silver, as seen in a specimen obtained by tracheal aspiration from a patient with the following history:

> The patient was an eight-month-old infant boy who had been the product of a 26-week pregnancy. The child was in respiratory distress from birth and was maintained in the intensive care nursery until he was two months of age. Following discharge the child continued to have respiratory problems and during one visit to the hospital a tracheal aspirate was obtained for cytologic study.

In contrast to some of the other organisms being discussed, *Histoplasma capsulatum* is more easily viewed in cytologic preparations utilizing special stains. In our laboratories we have found methenamine silver most useful for this purpose. The organism appears as a 1–5 micron round to oval body with budding. For diagnostic purposes it must be engulfed within macrophages or neutrophils.[4, 6, 7]

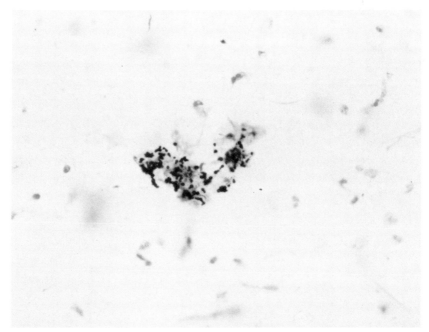

Fig. IV-39. Tracheal aspirate. *Histoplasma capsulatum.* These tiny, budding yeast-like organisms are completely contained within the cytoplasm of macrophages, staining only faintly in the background. Methenamine silver stain, × 400.

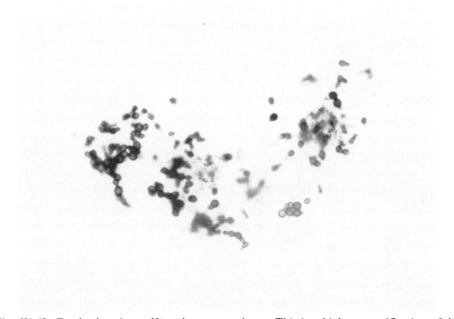

Fig. IV-40. Tracheal aspirate. *Histoplasma capsulatum.* This is a higher magnification of the organisms depicted in Figure IV-39. Methenamine silver stain, × 1000.

OPPORTUNISTIC MYCOTIC INFECTIONS

The organisms just discussed are primary pathogens and as such do not usually appear in pulmonary material in the absence of infection. Such a situation is not necessarily the case for the opportunistic fungi, which may also be observed in respiratory cytology. These latter organisms are usually considered saprophytes or contaminants, but may invade and produce infection in persons whose resistance has been lowered in some manner. The incidence of this type of infection has been increasing in recent years.[19] It is important that the presence of these fungi be noted so that appropriate further studies may be properly initiated to determine significant infection. Differential diagnosis of these organisms becomes more difficult than that of the yeast-like mycoses, as most of the former are characterized by branching or nonbranching hyphal fragments, with or without spores.

Aspergillus

Among the opportunists, we have observed various species of the genus aspergillus most frequently (Figs. IV-41 through IV-49). The very characteristic presentation is that of thick, uniform septate hyphae with 45°-angle, brush-like branching. The mycelial growth in pulmonary aspergillosis usually is not associated with the presence of conidiophores of fruiting heads, so that confusion with phycomycosis may occur. However, fungi producing the latter disease are not septate. The presence of septate-branching hyphae in cytologic material is strong morphologic evidence of infection.

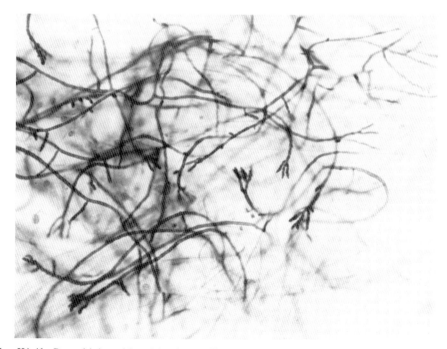

Fig. IV-41. Bronchial washing. Membrane filter preparation. Species of the genus Aspergillus. Papanicolaou stain, × 400.

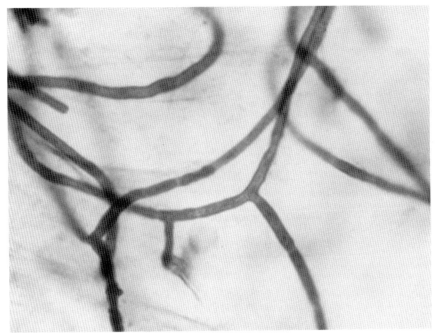

Fig. IV-42. Bronchial washing. Species of the genus Aspergillus. This is a higher magnification of the organism depicted in Figure IV-41. The hyphal fragments of Aspergillus are characterized by septation, uniformity, and 45° angle branching. Papanicolaou stain, × 1000.

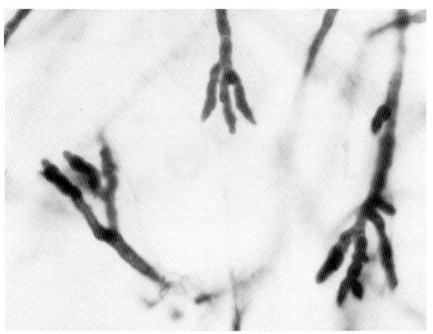

Fig. IV-43. Bronchial washing. Membrane filter preparation. Species of the genus Aspergillus. This is another field of the same specimen shown in Figures IV-41 and IV-42. The brush-like 45° angle branching is well shown here. Papanicolaou stain, × 1000.

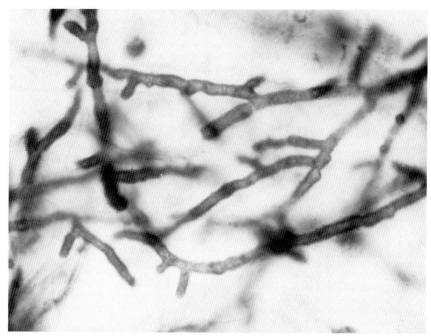

Fig. IV-44. Bronchial washing. Membrane filter preparation. Species of the genus Aspergillus. Papanicolaou stain, × 1000.

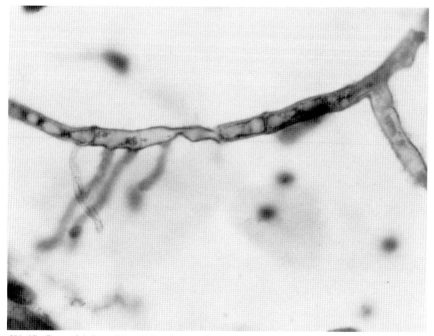

Fig. IV-45. Bronchial washing. Membrane filter preparation. Species of the genus Aspergillus. A branching septate hyphal fragment is shown, with some collapse of the cell wall. Papanicolaou stain, × 1000.

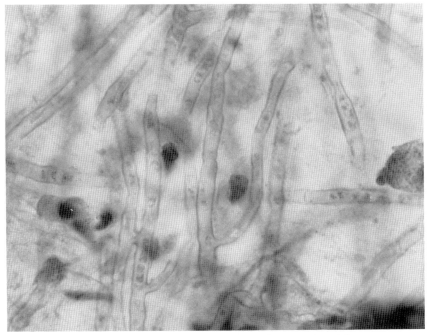

Fig. IV-46. Bronchial washing. Membrane filter preparation. Species of the genus Aspergillus. Papanicolaou stain, × 1000.

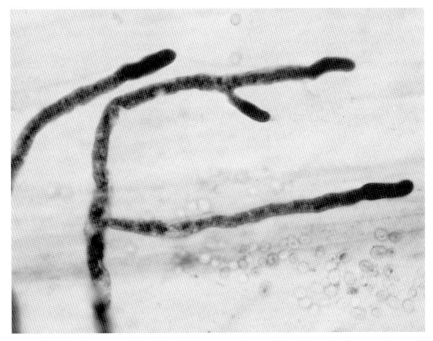

Fig. IV-47. Fresh sputum. Species of the genus Aspergillus. Papanicolaou stain, × 1000.

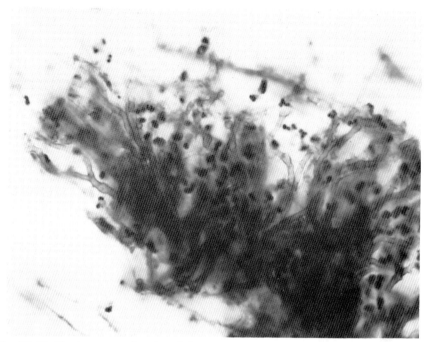

Fig. IV-48. Fresh sputum. Species of the genus Aspergillus. Papanicolaou stain, × 400.

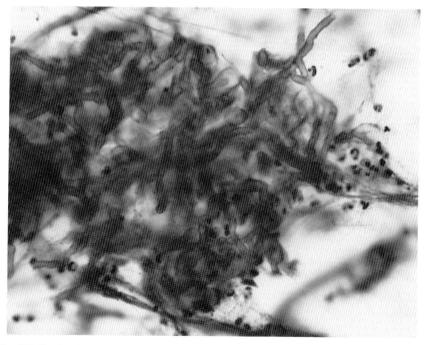

Fig. IV-49. Fresh sputum. Species of the genus Aspergillus. Papanicolaou stain, × 400.

Cultures alone may be positive in the absence of true infection. Occasionally sporeheads and mycelium may be seen.

Aspergillus, along with several other organisms, has been implicated in the production of cellular atypias easily mistaken for squamous-cell carcinoma.[6, 20-22] An example of such a problem is depicted in Figures IV-50 through IV-52. By x-ray it was seen that the patient had a cavitary lesion thought to be squamous-cell carcinoma. The patient in fact had a mycetoma. This case is discussed in further detail in Chapter V.

Occasionally large accumulations of crystals may be seen in association with the aspergillus hyphae. These are usually crystals of calcium oxalate (Figs. IV-53 and IV-54). Aspergillus species produce large quantities of oxalic acid, which is believed to precipitate in the tissue after chemical combination with tissue calcium compounds. This hypothesis has been well summarized by Nime and Hutchins.[23]

Phycomycetes and Phycomycosis

Phycomycosis has most recently been defined as an acute mycotic infection in which there is extensive inflammation and vascular thrombosis caused by invasion of the vessel walls and lumina by any one of a number of different organisms. The ever-growing list of mycoses capable of causing this very dangerous pathologic picture includes Absidia, Mucor, Rhizopus, and Basidiobolus.

The following clinical histories define the more common situations in which specimens from the respiratory tract may be submitted for cytologic interpretation.

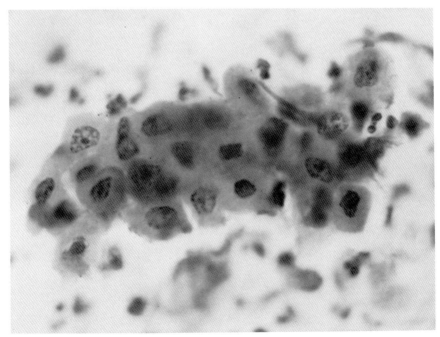

Fig. IV-50. Sputum prepared by Saccomanno technique. Atypical metaplastic cells from aspergilloma. Papanicolaou stain, × 1000.

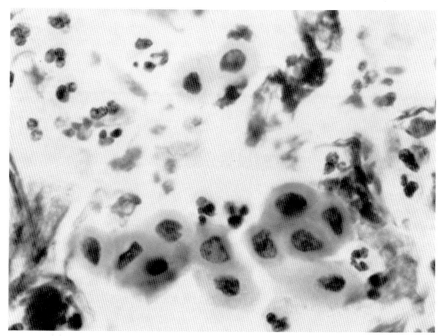

Fig. IV-51. Sputum prepared by Saccomanno technique. Atypical metaplastic cells from aspergilloma. This is an additional field from the case depicted in Figure IV-50. Papanicolaou stain, × 1000.

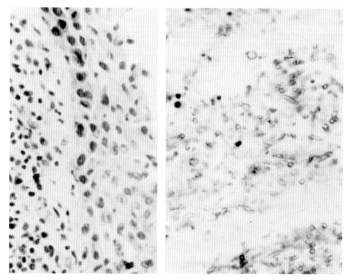

Fig. IV-52. Histologic section from aspergilloma. This is the tissue of origin of the cells shown in Figures IV-50 and IV-51. The histopathology of the lining of atypical metaplastic cells is shown in the left panel, and the mass of hyphae identified as species of the genus Aspergillus, which were present in the cavity, are shown in the right panel. Hematoxylin and eosin stain, × 250.

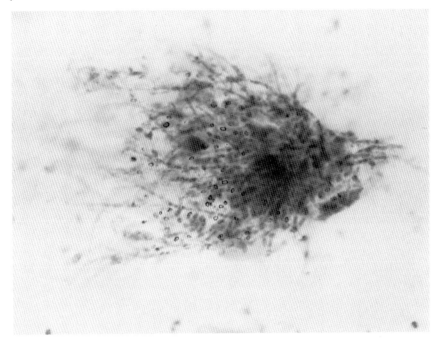

Fig. IV-53. Bronchial washing. Membrane filter preparation. Species of the genus Aspergillus with oxylate crystals. Papanicolaou stain, × 400.

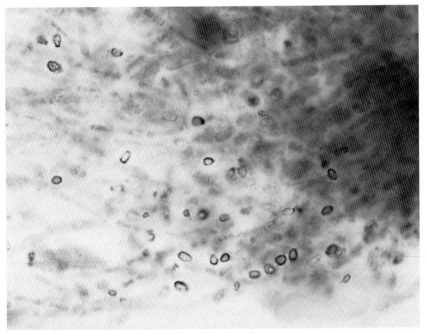

Fig. IV-54. Bronchial washing. Membrane filter preparation. Species of the genus Aspergillus with oxylate crystals. This is a higher magnification of the field shown in Figure IV-53. Papanicolaou stain, × 1000.

Case 1. A 62-year-old white man had an 11-year history of chronic lymphocytic leukemia. It had been under good control until three months prior to admission, when the leukemia was complicated by severe hemolytic anemia and increasing shortness of breath. Chest x-ray revealed multiple nodules scattered throughout both lungs, with questionable interstitial infiltrate in the lower lobes. He had a rapidly progressive pneumonia which did not respond to high-dose antibiotics. Sputum and bronchial washings were obtained with a request to look for pneumocystis.

Case 2. A 77-year-old woman with poorly controlled diabetes had a right upper-lobe pulmonary lesion described as irregular, about 6.0 cm in diameter, with a cavity. Bronchial brushings and transbronchial biopsy were performed. (Material from this case has been included with the kind permission of Dr. Donald Leonard, Cone Memorial Hospital, Greensboro, N. C.)

Case 3. A 45-year-old woman was found to have a mass in the left upper lobe. She was otherwise in good health with no underlying disease process. Bronchial washings were submitted for cytologic diagnosis.

In all three of these patients the correct diagnosis of phycomycosis was made first by cytologic study and later confirmed by biopsy.

Figure IV-55 illustrates those features of the hyphae, found in infected cases, that are useful in making the diagnosis of phycomycosis. As can be seen from the illustration, the fungus is characterized by ribbon-like, nonseptate coenocytic branching hyphae. They may vary greatly in width, from 6–50 microns. Culture is necessary to identify which fungus is causing the infection, as multiple organisms with identical tissue morphology may produce the disease.[21]

Materials from Cases 1 and 2 are shown in Figures IV-56 through IV-60. A very

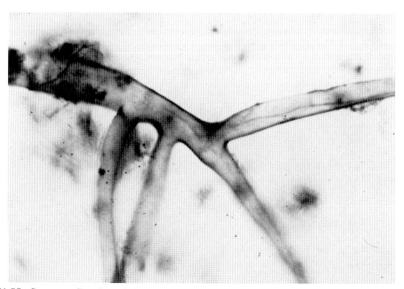

Fig. IV-55. Sputum. Species of the genus Rhizopus. This specimen is a simulation of patient material prepared by seeding mycelial fragments of Rhizopus from a fresh culture into a specimen of sputum. It shows particularly well the diagnostic features of the hyphae of the Phycomyces, which occur in the disease phycomycosis. These hyphae are ribbon-like, they vary markedly in their diameter, and they are nonseptate and coenocytic. Papanicolaou stain, × 1000.

Fig. IV-56. Bronchial brushing. Membrane filter preparation. Hyphal fragments from phycomycosis. Methenamine silver stain, × 400.

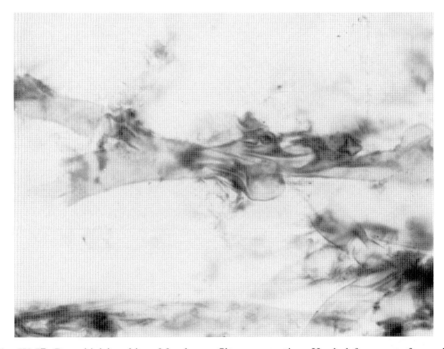

Fig. IV-57. Bronchial brushing. Membrane filter preparation. Hyphal fragments from phycomycosis. Papanicolaou stain, × 1000.

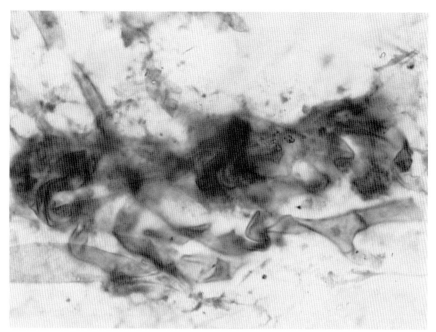

Fig. IV-58. Bronchial brushing. Hyphal fragments from phycomycosis. This is another field of the case depicted in Figure IV-57. Papanicolaou stain, × 1000.

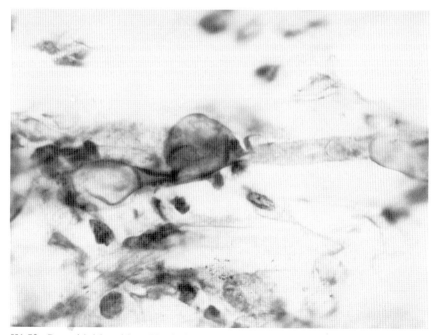

Fig. IV-59. Bronchial brushing. Hyphal fragments and sporangiophores from phycomycosis. This is another field of tissue from the patient depicted in Figures IV-57 and IV-58 and illustrates the rare phenomenon of sporangiophore formation in hyphal fragments associated with disease. Papanicolaou stain, × 1000.

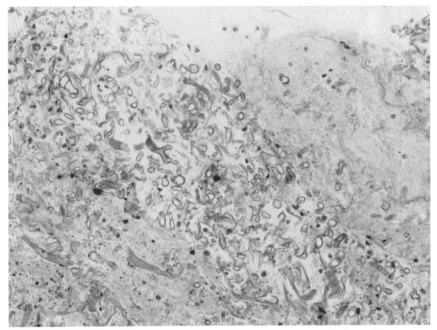

Fig. IV-60. Transbronchial needle biopsy of phycomycosis. Multiple branching, nonseptate hyphal fragments are present in a necrotic exudate. This is the tissue from the patient whose cytologic specimens are depicted in Figures IV-57–59. Hematoxylin and eosin stain, × 400.

unusual phenomenon of the production of sporangia in infected material is illustrated in Figure IV-59.

PARASITES

Pneumocystis carinii

In recent years the incidence of *Pneumocystis carinii* pneumonia has increased and is now recognized as potentially occurring in any situation of impaired immune response. More particularly it is seen in infants who are premature or debilitated, in immunologic disorders, in immunoglobulin defects, in the presence of therapy with corticosteroids and chemotherapy, and in renal transplants. The introduction of effective therapeutic drugs has markedly increased the clinical importance of ante-mortem diagnosis.[24]

Several years ago one of the authors (Johnston), in association with Dr. Edwin L. Kamstock, elected to study pneumocystis in laboratory mice so as to be able to correlate the appearance of the organism in tissue and respiratory secretions. Material from this study is illustrated in Figures IV-61 through IV-66. From this study it was learned that the eosinophilic exudate seen in the H&E stained sections of lung is also seen on the smears; that the large, active histiocytes seen in tissue are also present on smears; and that the morphology of the organism on smears is almost identical to that in sections.

On Papanicolaou-stained material, the organisms may be difficult to identify, as their staining is quite variable and faint, even in the most ideal of cases. Their most

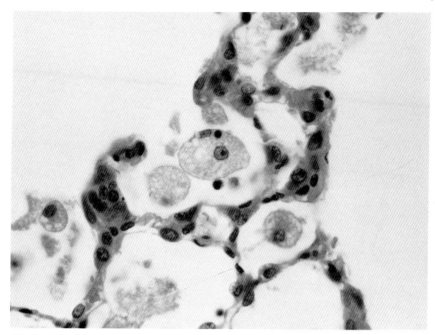

Fig. IV-61. Histologic section of mouse lung infected with *Pneumocystis carinii.* Hematoxylin and eosin stain, × 400.

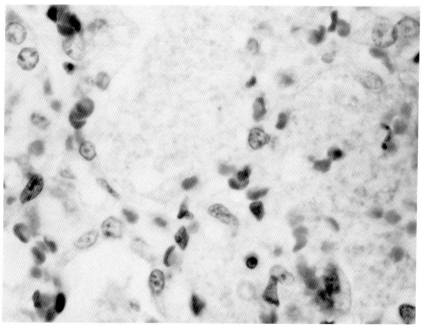

Fig. IV-62. Histologic section of mouse lung infected with *Pneumocystis carinii.* Embedded within the eosinophilic honeycombed exudate can be seen the faint outline of the cyst walls of the organism. Hematoxylin and eosin stain, × 1000.

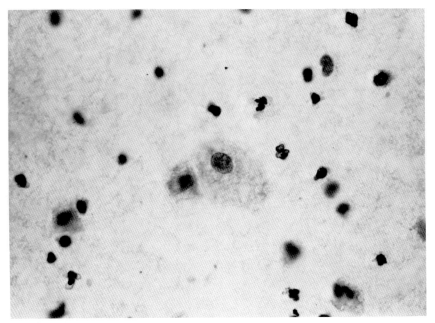

Fig. IV-63. Lung lavage. Membrane filter preparation. *Pneumocystis carinii* infection in mouse lung. Depicted in the center is one of the very large macrophages characteristically associated with this disease. Compare this morphology with that of the macrophages in Figure IV-61. Papanicolaou stain, × 400.

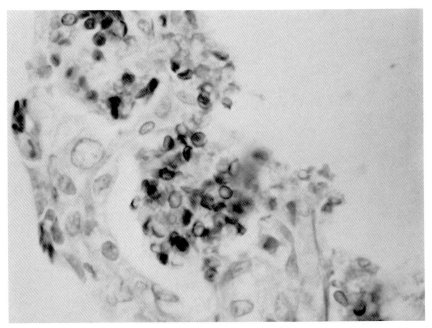

Fig. IV-64. Histologic section of mouse lung infected with *Pneumocystis carinii.* A large number of typical organisms is clustered together in the alveolar spaces. Methenamine silver stain, × 1000.

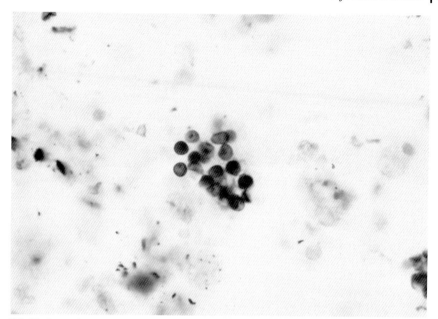

Fig. IV-65. Lung lavage. Membrane filter preparation. *Pneumocystis carinii.* Prepared from lavage material obtained from experimental *Pneumocystis carinii* infection in mice. Depicted are characteristic organisms as they appear on staining with methenamine silver. Methenamine silver stain, × 1000.

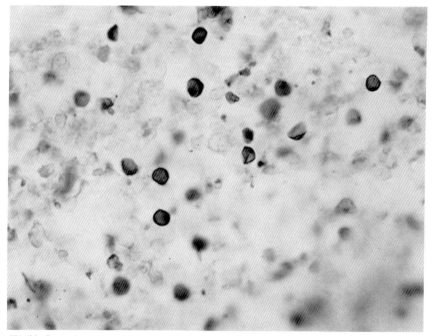

Fig. IV-66. Lung lavage. Membrane filter preparation. *Pneumocystis carinii* infection in mouse lung. This is another field from the specimen depicted in Figure IV-65. Methenamine silver stain, × 1000.

typical presentation on the Papanicolaou-stained smear is as a mass of partially eosinophilic or amorphous material. Within this mass may be a suggestion of small, superimposed circlets (Fig. IV-67, left panel, and Fig. IV-68). Although this presentation is not characteristic for pneumocystis, such a honeycombed, amorphous, eosinophilic mass should be further evaluated by special stains. In such a situation, one would decolorize this slide and restain with methenamine silver. This procedure immediately brings out the diagnostic features of these organisms (Figure IV-67, right panel). On methenamine silver stains, the organism is seen mainly as a spherical cyst measuring 6–8 microns in diameter, or approximately the size of an erythrocyte. Certain variations of this form can be seen. The organism can be cup shaped, crescent shaped, or crinkled. Depending upon the surface of the organism exposed to view, small globoid interior structures can be seen, some of which appear to be attached to the cyst wall (Fig. IV-65). Some laboratories prefer a Giemsa stain for identification. With this technique it is possible to identify up to eight structures occurring within the cyst. These structures are about 0.5–1.0 microns in diameter, are easily overlooked, and may be confused with granules or cell fragments. (See Chapter XI, Fig. XI-15.) Other organisms to be considered in the differential diagnosis are candida species and histoplasma (Fig. IV-69).

In response to requests from clinicians for more rapid diagnosis than is usually possible from the laboratory techniques already discussed, a number of laboratories have tried various quick-staining methods on air-dried smears. The Giemsa stain is appropriate for this. More recently Murad and associates have reported success in utilizing toluidine blue.[25]

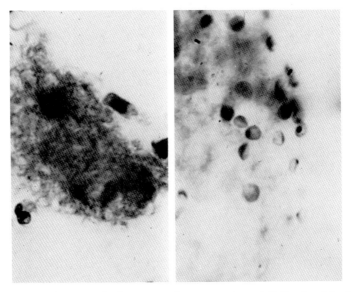

Fig. IV-67. Postbronchoscopy fresh sputum. *Pneumocystis carinii.* Depicted in the left panel is a mass of eosinophilic honeycombed exudate as stained by the Papanicolaou method. The right panel depicts the results of destaining and restaining this material with methenamine silver. The organisms are clearly shown. Left panel, Papanicolaou stain, × 400. Right panel, methenamine silver stain, × 1000.

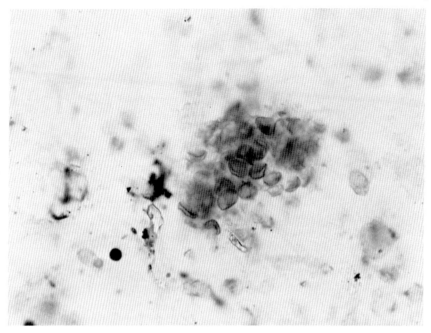

Fig. IV-68. Postbronchoscopy sputum. *Pneumocystis carinii.* Methenamine silver stain, × 1000.

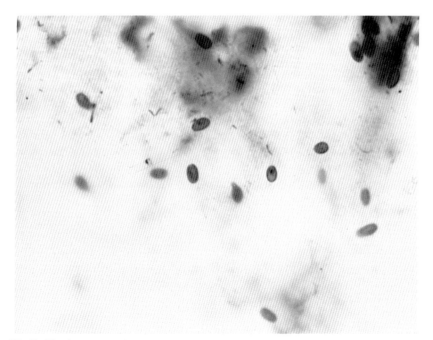

Fig. IV-69. Fresh sputum. Candida species. This figure depicts the characteristic morphology of Candida organisms when stained by the methenamine silver method. To the beginner this could be confused with the morphology of *Pneumocystis carinii.* Methenamine silver stain, × 1000.

Although in the literature one can find reports of success in diagnosing this organism in sputum, tracheal aspirates, and washings from hypopharynx, bronchus, and stomach,[26-28] we generally have been unsuccessful in finding the organisms in such preparations. We have seen the organism most frequently in transthoracic thin-needle aspirations, in bronchial brushing specimens, and in pulmonary lavage specimens. On only one occasion have we seen the organisms in sputum, and that sputum was a postbronchoscopy specimen.

Strongyloides stercoralis

During the last several years one of the authors (Johnston) has observed two pulmonary infections with the filariform larval stage of *Strongyloides stercoralis*. These case histories follow.

> **Case 1.** A 45-year-old man was seen in renal failure and received a cadaveric kidney transplant. Due to increasing creatinine on the third postoperative day, he was begun on high-dose steroids. Three weeks later he was readmitted with cough, dyspnea, abdominal pain, and bloody diarrhea. He developed pulmonary edema, and transtracheal aspirates revealed filariform larval forms of *Strongyloides stercoralis*. He was treated with thiabendazol, but had multiple cardiac arrythmias, renal failure, and died. Autopsy revealed extensive systemic strongyloidiasis.

> **Case 2.** A 77-year-old man had been treated with 40 mg per day prednisone for asthma at an outside hospital. He was referred to the pulmonary service at Duke University Medical Center and sputum cytologies were obtained. Both rhabditiform and filariform larvae of *Strongyloides stercoralis* were felt to be present in the specimens. The patient died two weeks later and was found at autopsy to have widely disseminated strongyloidiasis.

In both patients prior disease was existent. One patient was a renal transplant recipient who had been supported with heavy steroid dosage. The second patient had severe asthmatic bronchitis also treated with massive steroid therapy. In both situations the primary diagnosis was suggested by cytologic examination of sputum. The organisms observed measured 400–500 microns in length and exhibited closed gullets and slightly notched tails (Figures IV-70 through IV-74). This morphology differentiated them from *Ascaris lumbricoides*, *Necator americanus*, and *Ancylostoma duodenale*.[6, 21] One prior case of strongyloidiasis diagnosed in Papanicolaou-stained sputum smears has been reported by Kenney and Webber.[29]

Other Parasites

Other reports of the cytologic detection of parasitic pulmonary disease have included a report of pulmonary echinococcosis by Allen and Fullmer,[30] the lung fluke *Paragonimus kellicotti* by McCallum,[31] and the fluke *Paragonimus westermanii* by Willie and Snyder.[32]

VIRAL INFECTIONS

A number of laboratories have investigated the cellular changes in respiratory epithelium in response to a viral infection.[33-38] Particularly noteworthy here is the 1968 study of the cytological features of viral respiratory tract infections by Naib and associates.[36] Cellular changes were noted in 41 of 99 patients with culturally proven

Fig. IV-70. Fresh sputum. Filariform larva of *Strongyloides stercoralis.* Methenamine silver stain, × 1000.

Fig. IV-71. Fresh sputum. Filariform larva of *Strongyloides stercoralis.* At the extreme left margin of the figure is the tail with its blunt, slightly notched appearance. Methenamine silver stain, × 1000.

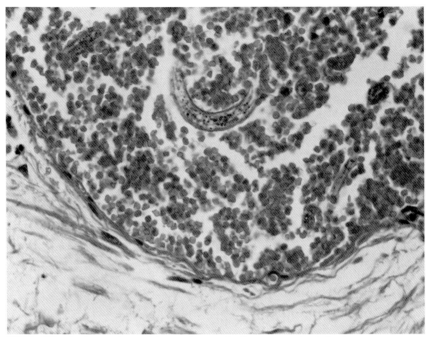

Fig. IV-72. Histologic section of lung. *Strongyloides stercoralis* infection. A portion of a filariform larva is seen within the lumen of a pulmonary vessel. Hematoxylin and eosin stain, × 400.

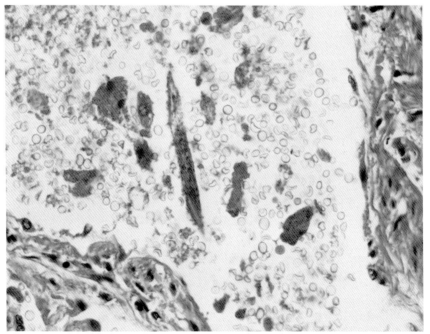

Fig. IV-73. Histologic section of lung. *Strongyloides stercoralis* infection. Portions of a filariform larva of *Strongyloides stercoralis* are observed in an alveolar space, mixed with blood and necrotic exudate. Hematoxylin and eosin stain, × 400.

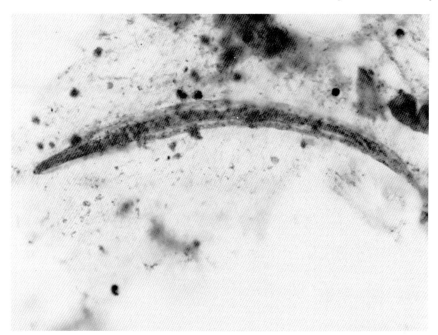

Fig. IV-74. Fresh sputum. Filariform larva of *Strongyloides stercoralis.* Papanicolaou stain, × 1000.

viral infection. The specificity of the cellular changes was confirmed in 76% of the cases by conventional virological procedures. Characteristic cellular changes were observed in patients with parainfluenza virus, adenovirus, and cytomegalovirus infections. The cellular changes in such patients can be divided into three general categories. First is that of a nonspecific cellular alteration characterized by ciliocytophthoria, first described by Papanicolaou in 1956.[39] This is a peculiar degeneration of the ciliated respiratory epithelium in which a pinching off occurs between the cilia-bearing cytoplasm and the nucleated cytoplasm, leaving an anucleated mass of cytoplasm bearing cilia and a degenerating nucleus and cytoplasm.[40, 41] An additional morphologic feature may be the presence of small, round eosinophilic masses in the cytoplasm. Examples of ciliocytophthoria are shown in Figures IV-75 through IV-78.

A more problematical type of nonspecific alteration is that of regeneration and repair atypia of the respiratory epithelium. This may present in sputum and bronchial material as tissue fragments composed of cells bearing enlarged hyperchromatic nuclei and enlarged prominent nucleoli.[20] A false-positive diagnosis of cancer is possible if there is no awareness of these changes or of a possible etiology. The tightly coherent feature of the cells in the tissue fragments, as well as the absence of atypical cells lying singly, are aids in avoiding a cancer diagnosis (Fig. IV-79).

The more specific and diagnostic cellular alterations are of the most widely applied, practical significance in the presence of infections with *Herpes simplex* and the cytomegalovirus. Changes with both of these infections are highly characteristic and have been well described in the literature.[34–38, 42–44] The hallmark of cellular alteration produced by *Herpes simplex* is that of cells with multiple, molded nuclei, which may contain eosinophilic, irregular inclusion bodies (Figs. IV-80 through IV-82), or which

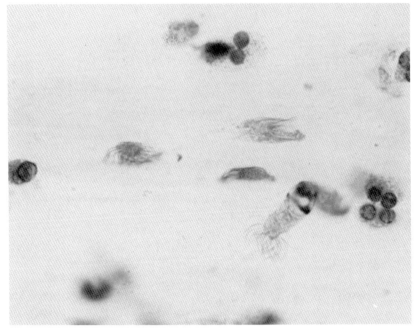

Fig. IV-75. Fresh sputum. Ciliocytophthoria. Papanicolaou stain, × 1000.

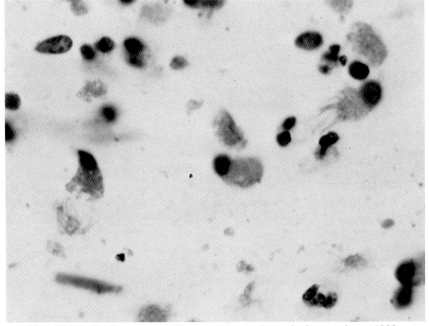

Fig. IV-76. Fresh sputum. Ciliocytophthoria. Papanicolaou stain, × 1000.

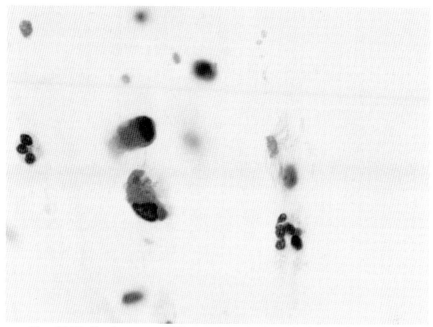

Fig. IV-77. Fresh sputum. Ciliocytophthoria. Papanicolaou stain, × 1000.

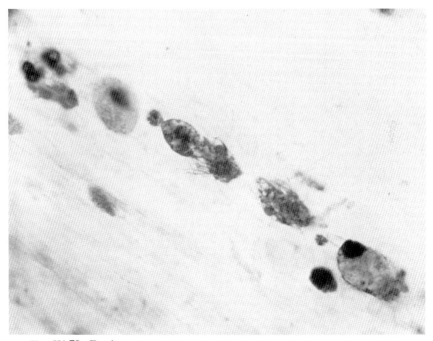

Fig. IV-78. Fresh sputum. Ciliocytophthoria. Papanicolaou stain, × 1000.

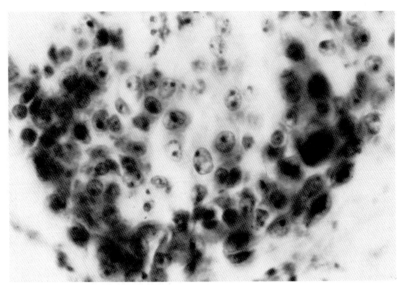

Fig. IV-79. Fresh sputum. Reactive bronchial epithelium from viral pneumonia. Papanicolaou stain, × 400.

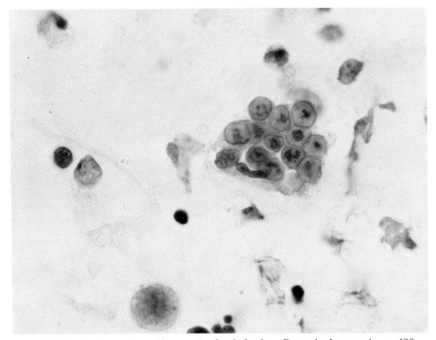

Fig. IV-80. Fresh sputum. *Herpes simplex* infection. Papanicolaou stain, × 400.

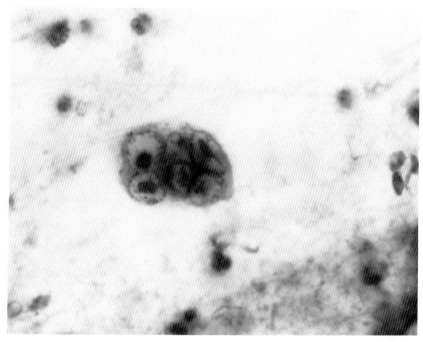

Fig. IV-81. Fresh sputum. *Herpes simplex* infection. Papanicolaou stain, × 1000.

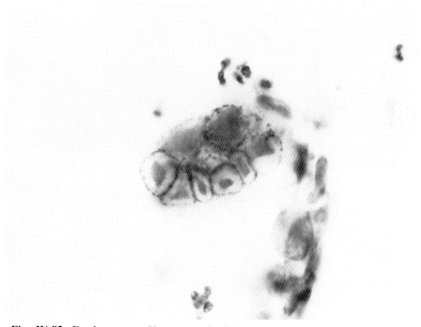

Fig. IV-82. Fresh sputum. *Herpes simplex* infection. Papanicolaou stain, × 1000.

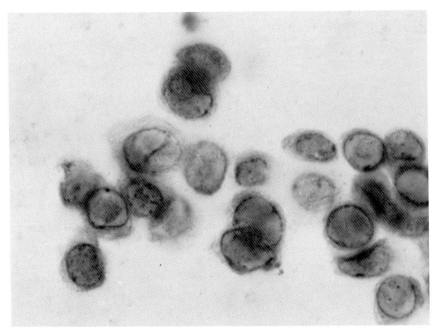

Fig. IV-83. Fresh sputum. *Herpes simplex* infection. Papanicolaou stain, × 1000.

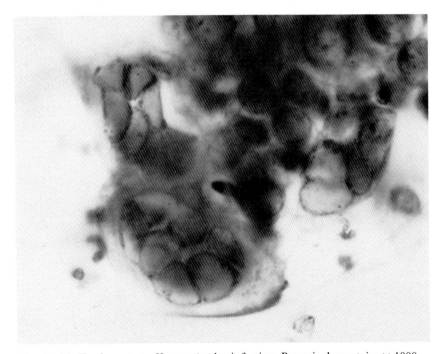

Fig. IV-84. Fresh sputum. *Herpes simplex* infection. Papanicolaou stain, × 1000.

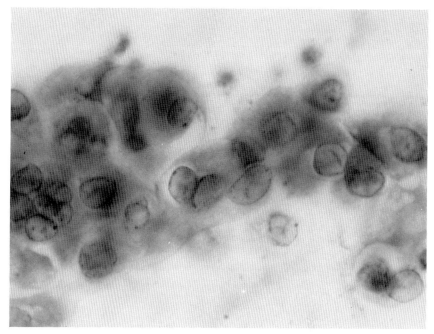

Fig. IV-85. Fresh sputum. *Herpes simplex* infection. Papanicolaou stain, × 1000.

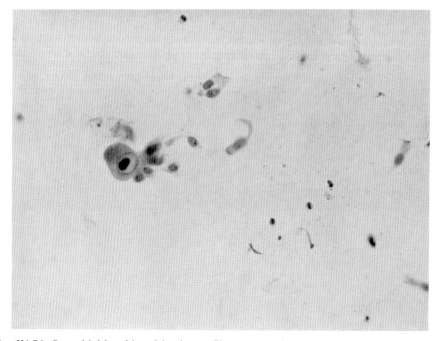

Fig. IV-86. Bronchial brushing. Membrane filter preparation. Cytomegalic inclusion disease in renal transplant recipient. Papanicolaou stain, × 400.

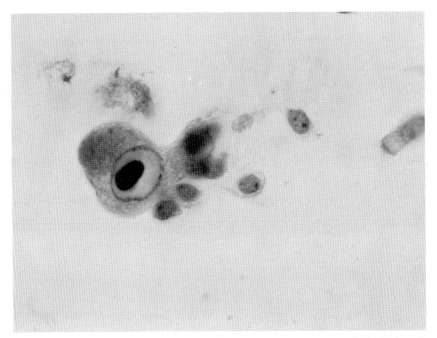

Fig. IV-87. Bronchial brushing. Membrane filter preparation. Cytomegalic inclusion disease. This is a higher magnification of the cell depicted in Figure IV-86. Note the rounded densities in the cytoplasm, which are intracytoplasmic inclusions from the virus. Papanicolaou stain, × 1000.

may exhibit a peculiar type of nuclear degeneration that appears as slate-gray homogenized nuclear contents (Figs. IV-83 through IV-85). Cells infected by the cytomegalovirus are larger and may show some multinucleation, but have fewer nuclei and none of the molding seen in *Herpes simplex*. Large basophilic, smooth intranuclear inclusions surrounded by a vary prominent halo and marked margination of chromatin on the inner surface of the nuclear membrane are present. Cytoplasmic inclusions also present in this disease may be manifested by the textured or "hammered aluminum" appearance of the cytoplasm (Figs. IV-86 and IV-87).

REFERENCES

1. Naib, Z. M.: Exfoliative cytology in fungus diseases of the lung. *Acta Cytol, 6:* 413–416, 1962.
2. Conant, N. F., Smith D. T., Baker, R. D., and Callaway J. M.: *Manual of Clinical Mycology*, 3rd ed. Saunders, Philadelphia, 1971.
3. Johnston, W. W., Schlein, B., and Amatulli, J.: Cytopathologic diagnosis of fungus infections. I. A method for the preparation of simulated cytopathologic material for the teaching of fungus morphology in cytology specimens. II. The presence of fungus in clinical material. *Acta Cytol, 13:* 488–495, 1969.
4. Johnston, W. W.: The cytopathology of mycotic infections. *Lab Med, 2:* 34–40, 1971.
5. Larsh, H. W. and Goodman, N. L.: Sputum Mycology. In: *Sputum, Fundamentals and Clinical Pathology.* MJ Dulfano, ed. Charles C Thomas, Springfield, Ill, 1973, pp 292–331.
6. Frable, W. J. and Johnston, W. W.: *Respiratory Cytopathology.* Tutorials of Cytology, Chicago, 1974.
7. Johnston, W. W. and Frable, W. J.: Cytopathology of the respiratory tract. *Am J Pathol 84:* 372–424, 1976.

8. Chipps, H. D. and Kraul, L. H.: Cytologic alterations in pulmonary tuberculosis which simulate carcinoma. *Cancer Res 10:* 210, 1950.

9. Palva, T. and Saloheimo, M.: Observations on the cytologic pattern of bronchial aspirates in pulmonary tuberculosis. *Acta Tuberc Scand 31:* 278–288, 1955.

10. Garret, M.: Cellular atypias in sputum and bronchial secretions associated with tuberculosis and bronchiectasis. *Am J Clin Pathol 34:* 237–246, 1960.

11. Roger, V., Nasiell, M., Nasiell, K., Hjerpe, A., Enstad, I., and Bisther, A.: Cytologic findings indicating pulmonary tuberculosis. II. The occurrence in sputum of epithelioid cells and multinucleated giant cells in pulmonary tuberculosis, chronic non-tuberculous inflammatory lung disease and bronchogenic carcinoma. *Acta Cytol, 16:* 538–542, 1972.

12. Nasiell, M., Roger, V., Nasiell, K., Enstaf, I., Vogel, B., and Bisther, A.: Cytologic findings indicating pulmonary tuberculosis. I. The diagnostic significance of epithelioid cells and Langhans giant cells found in sputum or bronchial secretions. *Acta Cytol 16:* 146 151, 1972.

13. Johnston, W. W. and Amatulli, J.: The role of cytology in the primary diagnosis of North American blastomycosis. *Acta Cytol 14:* 200–204, 1970.

14. Daniel, W. C., Nair, S. V., and Bluestein, J.: Light and electron microscopical observations of *Blastomyces dermatitidis* in sputum. *Acta Cytol:* In press.

15. Prolla, J. C., Rosa, U. W., and Xavier, R. G.: The detection of *Cryptococcus neoformans* in sputum cytology. Report of one case. *Acta Cytol 14:* 87–91, 1970.

16. Whitaker, D. and Sterrett, G. F.: *Cryptococcus neoformans* diagnosed by fine needle aspiration cytology of the lung. *Acta Cytol 20:* 105–106, 1976.

17. Johnston, W. W., Frable, W. J., Naylor, B., and Patten, S. F., Jr.: Diagnostic cytology seminar. *Acta Cytol 20:* 410–445, 1976.

18. Guglietti, L. C. and Reingold, I. M.: The detection of *Coccidioides immitis* in pulmonary cytology. *Acta Cytol 12:* 332–334, 1968.

19. Hutter, R. V. P. and Collins, H. S.: The occurrence of opportunistic fungus infections in a cancer hospital. *Lab Invest 11:* 1035–1045, 1962.

20. Koss, L. G.: *Diagnostic Cytology and Its Histopathologic Bases.* JB Lippincott, Philadelphia, 1968.

21. Johnston, W. W.: The cytopathology of mycotic and other infections. In *Compendium on Cytology.* Tutorials of Cytology, Chicago, 1976.

22. Louria, D. B., Lieberman, P. H., Collins, H. S., and Blevins, A.: Pulmonary mycetoma due to *Allescheria boydii. Arch Intern Med 117:* 748–751, 1966.

23. Nime, F. A. and Hutchins, G. M.: Oxalosis caused by Aspergillus infection. *Johns Hopkins Med J 133:* 183–194, 1973.

24. Rosen, P. P., Martini, N., and Armstrong, D.: *Pneumocystis carinii* pneumonia. Diagnosis by lung biopsy. *Am J Med 58:* 794–802, 1975.

25. Pritchett, P. S., Murad, T. M., and Webster, M. J.: Identification of *Pneumocystis carinii* infection in cytology specimens (abstract). *Acta Cytol 21:* 784–785, 1977.

26. Fortuny, I. E., Tempero, K. F., and Amsden, T. W.: *Pneumocystis carinii* pneumonia diagnosed from sputum and successfully treated with pentamidine isethionate. *Cancer 26:* 911–913, 1970.

27. Repscher, L. H., Schroter, G., and Hammond, W. S.: Diagnosis of *Pneumocystis carinii* pneumonitis by means of endobronchial brush biopsy. *N Engl J Med 287:* 340–341, 1972.

28. Kim, H. and Hughs, W. T.: Comparison of methods for identification of *Pneumocystis carinii* in pulmonary aspirates. *Am J Clin Pathol 60:* 462–466, 1973.

29. Kenney, M. and Webber, C. A.: Diagnosis of strongyloidiasis in Papanicolaou-stained sputum smears. *Acta Cytol 18:* 270–273, 1974.

30. Allen, A. R. and Fullmer, C. D.: Primary diagnosis of pulmonary echinococcosis by the cytologic technique. *Acta Cytol 16:* 212–216, 1972.

31. McCallum, S. M.: Ova of the lung fluke *Paragonimus kellicotti* in fluid from a cyst. *Acta Cytol 19:* 279–280, 1975.

32. Willie, S. M. and Snyder, R. N.: The identification of *Paragonimus Westermanii* in bronchial washings. Case report. *Acta Cytol 21:* 101–102, 1977.

33. Beale, A. J. and Campbell, W. A.: A rapid cytological method for the diagnosis of measles. *J Clin Pathol 12:* 335–337, 1959.

34. Koprowska, I.: Intranuclear inclusion bodies in smears of respiratory secretions. *Acta Cytol 5:* 219–228, 1961.

35. Warner, N. E., McGrew, E. A., and Nanos, S.: Cytologic study of the sputum in cytomegalic inclusion disease. *Acta Cytol 8:* 311–315, 1964.

36. Naib, Z. M., Stewart, J. A., Dowdle, W. R., Casey, H. L., Marine, W. M., and Nahmias, A. J.: Cytological features of viral respiratory tract infections. *Acta Cytol 12:* 162–171, 1968.

37. Nash, G. and Foley, F. D.: Herpetic infection of the middle and lower respiratory tract. *Amer J Clin Pathol 54:* 857–863, 1970.
38. Jain, U., Mani, K., and Frable, W. J.: Cytomegalic inclusion disease: Cytologic diagnosis from bronchial brushing material. *Acta Cytol 17:* 467–468, 1973.
39. Papanicolaou, G. N.: Degenerative changes in ciliated cells exfoliating from the bronchial epithelium as a cytologic criterion in the diagnosis of diseases of the lung. *NY State J Med 56:* 2647–2650, 1956.
40. Pierce, C. H. and Hirsch, J. G.: Ciliocytophthoria: Relationship to viral respiratory infections of humans. *Proc Soc Exp Biol Med 98:* 489–492, 1958.
41. Pierce, C. H. and Knox, A. W.: Ciliocytophthoria in sputum from patients with adenovirus infections. *Proc Soc Exp Biol Med 104:* 492–295, 1960.
42. Frable, W. J., Frable, M. A., and Seney, F. D.: Viral infections of the respiratory tract. Cytopathologic and clinical analysis. *Acta Cytol 21:* 32–36, 1977.
43. Frable, W. J. and Kay, S.: Herpes virus infection of the respiratory tract. Electron microscopic observation of the virus in cells obtained from a sputum cytology. *Acta Cytol 21:* 391–393, 1977.
44. An-Foraker, S. H. and Haesaert, S.: Cytomegalic virus inclusion body in bronchial brushing material. *Acta Cytol 21:* 181–182, 1977.

Epidermoid Carcinoma, Keratinizing and Nonkeratinizing

CHAPTER

V

Probably no phenomenon in medicine has been as dramatic as the rise in incidence of lung cancer in the last 30 years. Although there is some evidence that the rapid rise of the number of cases of this cancer in males may be leveling off, the death rate from lung cancer each year makes this a major health problem for men.[1] It is the number 1 killing cancer in males (estimated deaths 89,000 for 1977), and, because of the recent alarming increase in lung cancer in women, it now ranks third as the cause of cancer deaths in females.[2-4] Both incidence curves parallel the remarkable rise in smoking, first among males after World War I, and then in females following World War II.[2] Heavy cigarette smoking (two packs per day or more) correlates extremely well with incidence of lung cancer at all ages in adults, having an average lag period, between the time smoking is begun and the onset of clinical cancer of about 25 years.[5-7] Environmental pollution and increasing urbanization have also been shown to correlate with lung cancer data, but less clearly than cigarette smoking.[8]

Epidermoid carcinomas (both keratinizing and nonkeratinizing types) of the lung appear clinically as neoplasms of major bronchi. These two varieties account for approximately 50% of lung cancers in males and 20% in females.[3] Recent studies in connection with early lung cancer screening projects suggest that this tumor actually begins in a secondary and tertiary bronchus in many cases, growing to a size in its evolution that makes the neoplasm seem to have arisen from a more proximal bronchus.[9] Its clinical course is one of local infiltration followed shortly by lymph node and distant metastases.[10] The vascularity of the lung allows the tumor early access to blood vessels for dissemination, and invasion of vascular channels is usually easy to demonstrate histologically in the primary neoplasm. Epidermoid cancer's symptomatic stages are but a brief moment in their total life history, having followed development from areas of squamous metaplasia, basal-cell hyperplasia, dysplasia, and carcinoma *in situ* until invasion occurs and clinical symptoms are produced.[11, 12] The progression of events, seen in both man and experimental animals as the result of exposure to cigarette smoke or in the latter to both cigarette smoke and known carcinogens, is estimated to take an average of 10 or more years.[13, 14, 15] Thus the potential for early cytologic detection of this neoplasm appears possible and efforts in this respect have shown some success.[8]

The usual clinical situation is the patient with a symptomatic lung mass or infiltrate from whom sputum is sent for cytologic examination. Prognostically, for 100 such patients with proven lung cancer, half upon first examination will be found untreatable for cure because of documented distant disease or medical contraindications.[10] Of the 50 remaining patients who may undergo thoracotomy for surgical cure, only 25 will be resectable. Of those 25 whose tumor is resectable, only 12 by the most optimistic forecast may be expected to live five years. Cytologic diagnosis, while accurate, has not contributed to any improvement in these survival rates.[16]

The authors prefer the histologic nomenclature of the World Health Organization classification of lung cancer for cytologic reporting rather than a categorical or numerical classification.[17] It is realized that a diagnosis on every submitted sample may not necessarily be unequivocal using this type of reporting. A diagnosis may be modified to indicate that the cytology is suspicious for a lung cancer or that the cells present are only of dysplastic type. Terms describing the background of the smear (necrotic, inflammatory, etc.), as well as other cell types present, and their degree of atypia are also used. Recommendations for additional specimens are important. Using this type of terminology the cytopathologist must examine cellular constituents and characteristics of the smear carefully and weigh these findings against a set of standard criteria for each lung cancer type before rendering a final diagnosis. It must also be realized that from cytology alone it is not possible to recognize every tumor type within the World Health Organization classification. The authors believe that it is possible to diagnose accurately from cytology alone the major categories: epidermoid carcinoma with keratinization and without keratinization, large-cell undifferentiated carcinoma and its subtype giant-cell carcinoma, small-cell undifferentiated carcinoma, and adenocarcinoma. With some experience it is also possible to recognize three types of adenocarcinoma: bronchogenic acinar, bronchogenic papillary, and terminal bronchiolo-alveolar cell carcinoma. Although in theory it should be possible to recognize the combined epidermoid carcinoma and adenocarcinoma, the authors have found this difficult.

EPIDERMOID CARCINOMA WITH KERATINIZATION

The cytologic criteria for this tumor type are presented in Table V-1. Cells found in respiratory specimens are obviously malignant in this type of carcinoma of the lung. Most important is identification of large cells with aberrant shapes of both the cytoplasm and nucleus. Preservation of these cells is usually satisfactory enough so that chromatin detail reflects its abnormal distribution in the nucleus.

TABLE V-1
Cytologic Criteria: Epidermoid Carcinoma With Keratinization

Large single squamous cells
Aberrant shapes of cell and nucleus
Intense cytoplasmic eosinophilia and orangeophilia
Irregular chromatin distribution
Hyperchromasia of the nucleus

Hyperchromasia alone should not be used as a criterion for malignancy, since this may reflect only degeneration of the nucleus. The cytoplasm is usually extremely dense and its staining is intensely orangeophilic. The irregular lines of keratinization usually seen within the cytoplasm may be accentuated by lowering the substage condenser of the microscope. Many of the tumor cells have a distinct ectoplasmic border. Several examples of cells from typical cases of keratinizing epidermoid carcinoma are presented in Figures V-1, V-2, and V-3. There are at least a few of these unequivocally malignant cells in each smear, and at least two such cells should be seen on two different slides before the diagnosis is made. The keratinizing squamous tumor cells are frequently accompanied by small dysplastic keratinized cells in groups or sheets (Fig. V-4). Some tumor cells may not contain nuclei (squamous ghosts). Dysplastic squamous cells and/or squamous ghosts may be the first evidence of malignancy encountered in screening a respiratory specimen for lung cancer. They are not unequivocally diagnostic of keratinizing epidermoid carcinoma, but they strongly suggest that possibility. If no typical malignant cells are found, a request for additional specimens is advised.

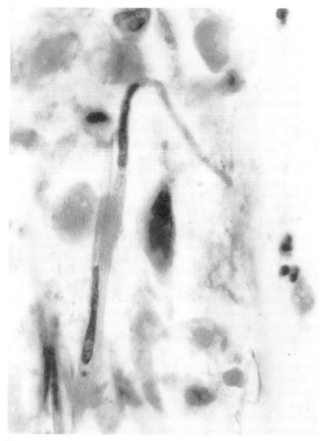

Fig. V-1. Fresh sputum, keratinizing epidermoid carcinoma. Pleomorphic cells with dense cytoplasm and irregular nuclear shape. Keratin masses (squamous ghosts) are also present. Papanicolaou stain, × 1000.

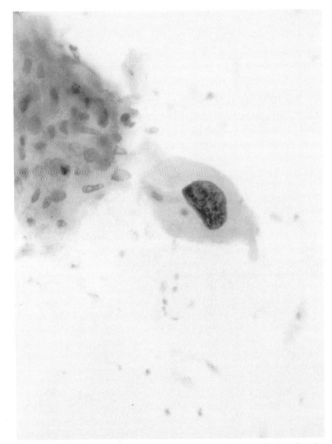

Fig. V-2. Sputum, keratinizing epidermoid carcinoma. Sheet of dysplastic squamous cells with pale nuclei. One large epidermoid carcinoma cell with irregular distribution of nuclear chromatin. Dysplastic cells frequently accompany typical keratinized epidermoid carcinoma cells. Papanicolaou stain, × 1000.

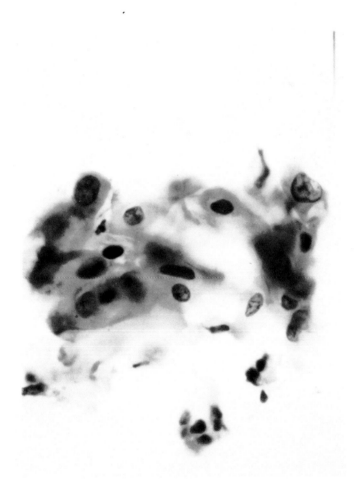

Fig. V-3. Sputum, keratinizing epidermoid carcinoma. Sheet of dysplastic and malignant cells from keratinizing epidermoid carcinoma. Papanicolaou stain, × 600.

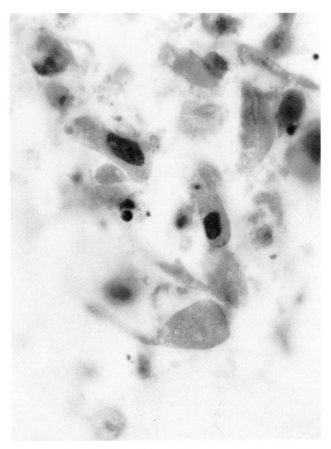

Fig. V-4. Fresh sputum, keratinizing epidermoid carcinoma. Pleomorphic, somewhat degenerated, malignant cells in a field, with several squamous ghosts. Papanicolaou stain, × 1000.

Careful examination of sputum smears from keratinizing epidermoid carcinoma will frequently reveal a faint granular background or tumor diathesis. Although this is not as reliable a criterion for malignant tumors in respiratory cytopathology as it is in cervical-vaginal cytopathology, in screening respiratory material it may still be a valuable first clue that a malignant neoplasm is present. This is particularly true if the diathesis occurs with little or no evidence of inflammation. If extensive inflammatory exudate is present, even with abnormal squamous cells, caution should be exercised in making a diagnosis of keratinizing epidermoid carcinoma. Metaplastic responses with degenerative atypia of the metaplastic squamous cells may be seen frequently in pneumonia or other nonneoplastic reactions of the respiratory tract.

Both bronchial brushings and bronchial washings from keratinizing epidermoid carcinoma provide cells identical to those seen in sputum. There may be more of them depending upon how close to the tumor the bronchoscope is at the time of sampling. Cells from a bronchial brushing specimen are seen in Figure V-5. Both a single bizarre cell and several dysplastic squamous cells are present. Occasionally the brushing will pick up a sheet of less differentiated cells (Fig. V-6) in an otherwise

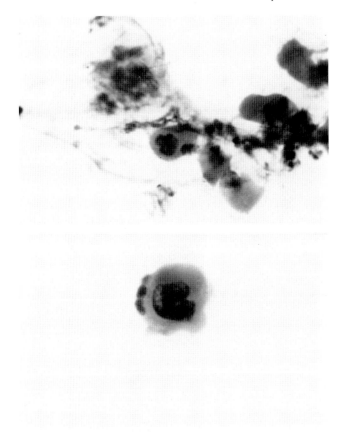

Fig. V-5. Bronchial brushing, keratinizing epidermoid carcinoma. Pleomorphic single tumor cell (bottom panel) and smaller dysplastic cells similar to those seen in sputum and bronchial washings of keratinizing epidermoid carcinoma. Papanicolaou stain, × 600.

typical keratinizing epidermoid carcinoma. A few of these cells have nucleoli and the cytoplasm is less dense; it may also be amphophilic or basophilic. If the majority of the tumor cells exhibit keratinization, the histology of the tumor will be a keratinizing epidermoid carcinoma even if a few less differentiated cells are seen in a brushing specimen.

EPIDERMOID CARCINOMA WITHOUT KERATINIZATION

Table V-2 lists the authors' cytologic criteria for this tumor type. There may be some variation in the cytology of this neoplasm, depending upon the type of specimen submitted and its method of preparation. Large bizarre and aberrantly shaped cells are similar to keratinizing epidermoid carcinoma seen in sputum. The cytoplasm is either basophilic or amphophilic. The cells can be identified as squamous from close examination of the cytoplasmic border, which will appear prominent or double. Examples of these cells are seen in Figures V-7 and V-8. Nuclear detail is not evident in the cell cluster in Figure V-7, but the prominent ectoplasmic border is obvious.

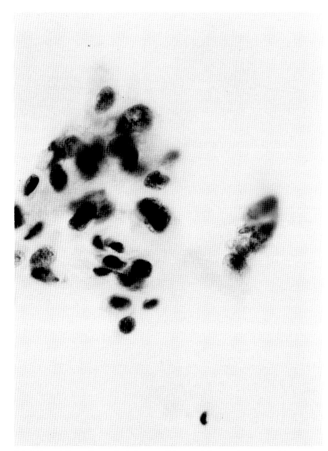

Fig. V-6. Bronchial brushing, keratinizing epidermoid carcinoma. Single irregular sheet of less differentiated tumor cells that may occasionally be seen in this type of sample from a keratinizing epidermoid carcinoma. Papanicolaou stain, × 560.

TABLE V-2

Cytologic Criteria: Epidermoid Carcinoma Without Keratinization

Large single cells
Aberrant cell shapes
Basophilic-amphophilic cytoplasmic staining
Prominent cell borders (ectoplasm) in at least
    some cells
Hyperchromasia of the nucleus
Distinct nuclear chromatin
Nucleoli seen frequently in bronchial washing and
    bronchial brushing specimens
Less differentiated tumor cells seen in bronchial
    washing and bronchial brushing specimens

Despite the nucleolus in the single cell (Fig. V-7), the double cell border identifies it as squamous. As in keratinizing epidermoid carcinoma, in well-preserved smears nuclear chromatin is distinct and abnormally distributed. Nucleoli are more frequently seen in bronchial washing and brushing specimens. The cytoplasm may be degenerated in many of the tumor cells found in sputum smears (Fig. V-8), but the high nuclear cytoplasmic ratio and irregular nuclear border and chromatin distribution identify them as neoplastic cells. Usually enough cells with intact cell borders are present so that they can be identified as squamous, or a few keratinized dysplastic cells that may accompany the less differentiated tumor cells can be used for identification of the carcinoma as epidermoid.

Figures V-9 and V-10 illustrate cells from nonkeratinizing epidermoid carcinoma seen in bronchial washings. The cytoplasm is degenerated to the extent that the tumor type would not be identifiable unless some cells with dense cytoplasm or intact cell borders could be identified (Fig. V-7). The arrangement of the cells in Figure V-9 might suggest an adenocarcinoma. That diagnosis would be reinforced by the finely vacuolated cytoplasm. Note the flat border and definite space between the most

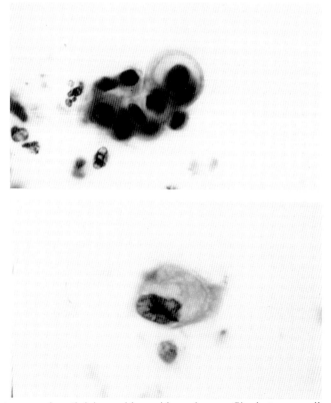

Fig. V-7. Sputum, nonkeratinizing epidermoid carcinoma. Single tumor cell with nucleolus (bottom panel) and cluster of tumor cells with variation in nuclear shape but only intense nuclear hyperchromasia. Note several cells with prominent ectoplasmic border, identifying the tumor as an epidermoid carcinoma. Papanicolaou stain, × 600.

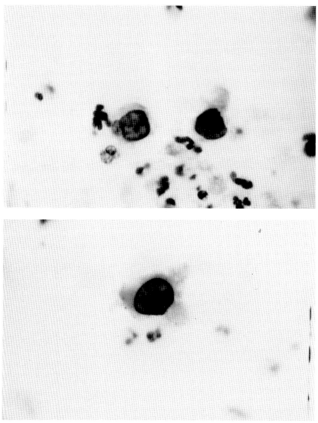

Fig. V-8. Sputum, nonkeratinizing epidermoid carcinoma. Single pleomorphic tumor cells with degeneration of the cytoplasm. Note irregular nuclear border and high nuclear cytoplasmic ratio. Cytoplasm stains basophilic. Other tumor cells with identifiable epidermoid features would have to be found to classify the neoplasm correctly from cytology only. Papanicolaou stain, × 600.

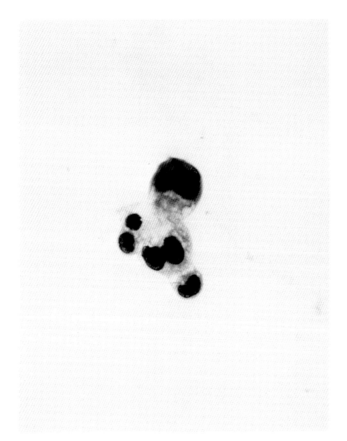

Fig. V-9. Bronchial washing, nonkeratinizing epidermoid carcinoma. Cluster of cells with finely vacuolated cytoplasm. Sharp border between cell cluster and small cell at the bottom of cluster and prominent ectoplasmic border of the largest cell identify the lesion as an epidermoid carcinoma. Papanicolaou stain, × 600.

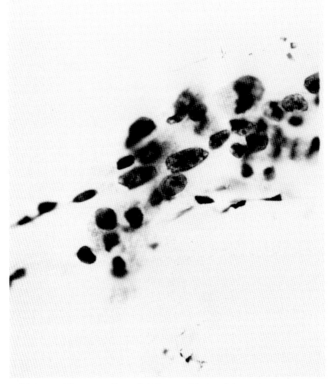

Fig. V-10. Bronchial washing, nonkeratinizing epidermoid carcinoma. Sheet of cells with irregular nuclei, but very degenerated cytoplasm. Elongated shape of the nuclei suggests an epidermoid carcinoma. Tumor cells with better preserved cytoplasm would have to be found for an exact identification of tumor type. Papanicolaou stain, × 600.

inferior cell and the adjacent cells in the group, as well as the double cell border along the margin of the large cell superior in position within the same group. Both features indicate this is an epidermoid tumor.

Bronchial brushings may also be confusing when sampling is such that practically all of the cells in the smear are malignant. Such a smear provides nothing that is normal for comparison. Because of the uniformity of the tumor cells (Figs. V-11 and V-12), the cytopathologist may begin to doubt that the cells are even malignant. Note the high nuclear cytoplasmic ratio and the distinct nuclear chromatin. Each cell has a prominent nucleolus. Even in these large sheets excellent preservation of the cells reveals the nuclear abnormalities are consistently present throughout the entire sheet. This will be true even if focusing of the microscope is required because of the size of the sheet or cluster of tumor cells. The brushing specimen in Figure V-12 does show clustering of tumor cells. A diagnosis of adenocarcinoma should not be made if the definite cell boundaries are seen as well as the sharp space between adjacent tumor cells in part of the cluster.

Epidermoid carcinoma may reach a large size and undergo central necrosis with formation of a cavity that is demonstrable radiographically. This is usually seen with heavily keratinized tumors and very rarely occurs with nonkeratinized carcinomas.

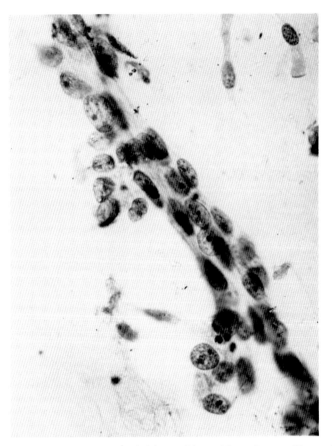

Fig. V-11. Bronchial brushing, nonkeratinizing epidermoid carcinoma. Sheet of cells similar to Figure V-10 but with better preservation of the cytoplasm and evidence of sharp cell borders. Polarity of the tumor cells is quite uniform, and the very high nuclear cytoplasmic ratio and distinct nuclear chromatin are the only important malignant features of these cells. Papanicolaou stain, × 600.

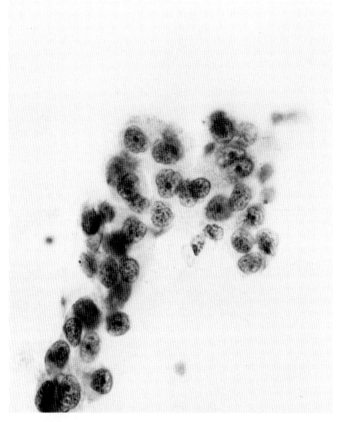

Fig. V-12. Bronchial brushing, nonkeratinizing epidermoid carcinoma. Cells in cluster arrangement that might suggest an adenocarcinoma. Depth of focus is minimal, and many of the cells have a prominent cytoplasmic border. Compare with brushings and washings from adenocarcinoma, Figures VIII-8, VIII-10, and VIII-25. Papanicolaou stain, × 600.

Sputum from the former cases may have a characteristic cytologic pattern of many small, parakeratotic dysplastic cells lying in a thick and uniform granular background (Fig. V-13). There will also be a few large, malignant tumor cells typical of keratinizing epidermoid carcinoma, and they should be sought for in confirming the diagnosis.

DIFFERENTIAL DIAGNOSIS

The respiratory tract responds to a large number of pathologic insults by a process of squamous metaplasia and varying degrees of atypia and dysplasia. Some of the dysplasias may be precancerous, and the separation of these from reactive atypias is one of the cytopathologist's most difficult tasks. Some of the cases discussed below illustrate the problems presented by dysplastic cells of the keratinized variety.

Figure V-14 illustrates isolated atypical squamous cells in sputum from a patient with an unresolved pneumonia. These cells have some variation in size and shape but possess badly degenerated nuclei. There are polymorphonuclear leukocytes

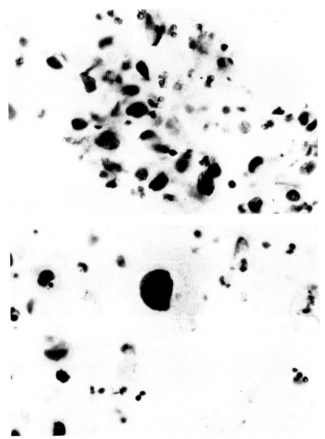

Fig. V-13. Sputum, epidermoid carcinoma, cavitating. Small, degenerated sheet of epidermoid carcinoma cells (upper panel) with granular debris in the background. Similar background in bottom panel with single degenerated large tumor cell having extremely hyperchromatic nucleus. Papanicolaou stain, × 600.

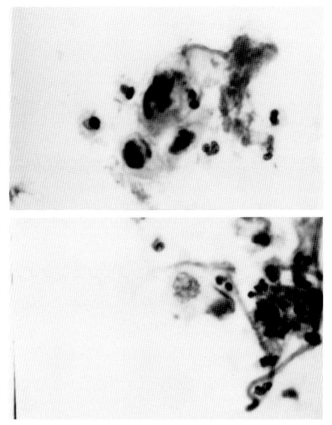

Fig. V-14. Endotracheal aspiration. Metaplasia and mild atypia of squamous cells in patient with pneumonia. There is both degeneration of the cytoplasm and nucleus, simulating dysplastic changes in squamous cells. Note inflammatory cells near or in the cytoplasm of these atypical cells. Papanicolaou stain, × 600.

around the cells and in the cytoplasm, an unusual finding in epidermoid carcinoma cells. The nuclei lack any definite nuclear chromatin abnormalities, and the cytoplasm is too degenerate for any diagnosis other than reactive atypia of squamous cells. Similar cells are quite common in specimens obtained by tracheal aspiration, either through a tracheostomy or by endotracheal intubation. The extent of this type of atypia and its differential diagnosis from carcinoma have been well described by Nunez et al.[18]

Figures V-15 and V-16 are from the sputum and bronchial washings of an elderly man with a large lung mass and a long history of rheumatoid arthritis. The cells are very abnormal in shape, though not large. The nuclei are very hyperchromatic but exhibit no nuclear chromatin detail. No cells with good preservation could be observed in the several specimens obtained. There is a marked background inflammation, but there is also necrotic debris in close approximation to the abnormal cells. The cells were considered dysplastic, but it was felt that carcinoma could not be entirely excluded. Biopsies could not confirm the diagnosis of carcinoma or a specific

Fig. V-15. Bronchial washing. Rheumatoid granuloma eroding bronchus. Degenerated atypical cells of uncertain type collected around necrotic debris. The cells probably represent examples of epitheloid cells that palisade at the periphery of such a granuloma. Papanicolaou stain, × 560.

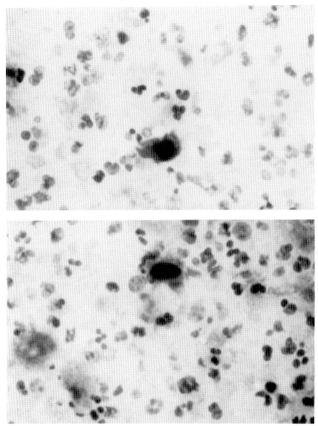

Fig. V-16. Bronchial washing. Rheumatoid granuloma eroding a bronchus. Same case as Figure V-15. Single degenerated atypical squamous cells, with marked hyperchromasia but no chromatin detail. An area of atypical squamous metaplasia adjacent to the area of erosion of the bronchus by the granuloma is the probable origin of these cells. Note the extensive inflammatory background. Papanicolaou stain, × 600.

pathologic process. Sometime later the patient expired. The mass in the lung was a large rheumatoid granuloma eroding into a segmental bronchus. There was adjacent squamous metaplasia. The cluster of cells and debris in Figure V-15 may represent part of the rheumatoid granuloma, since there is a slight suggestion of a palisaded arrangement of some of the elongated cells. The single dysplastic and degenerated cells are probably from the area of squamous metaplasia, which was focally atypical. The overall pattern of these smears might have suggested a cavitating epidermoid carcinoma, but that diagnosis could not be made without definite identification of tumor cells or the presence of large numbers of the degenerated dysplastic cells.

Figure V-17 demonstrates one of many sheets of dysplastic small squamous cells exfoliated in several sputums from a middle-aged man with a lung mass. Radiographically the mass appeared to be a lung cancer. These cells, though impressive in number for any given smear, are too small and uniform to represent a squamous-cell carcinoma. There are some nuclear abnormalities in chromatin distribution. The nuclear cytoplasmic ratio is consistent with that of an immature metaplastic cell.

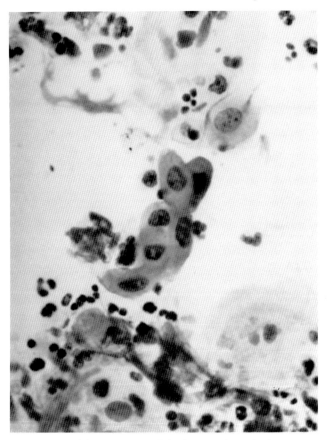

Fig. V-17. Sputum. Patient with mycetoma of the lung and atypical squamous metaplasia of the lining of the mycetoma cavity. Sheet of small, atypical squamous cells with some irregularity and hyperchromasia of nuclei. Note the inflammatory background. No typical keratinized epidermoid cancer cells were found. Fragments of hyphae of aspergillus were found upon review of several of the sputum smears. Papanicolaou stain, × 600.

Note the marked amount of inflammatory exudate. The mass was excised and proved to be a mycetoma of aspergillus filling a squamous-lined cavity (see Chapter IV). A section of the lining epithelium illustrates the probable source of the atypical squamous cells seen in the sputum (Fig. V-18). Review of the several sputum smears revealed a few fragments of hyphae that had been overlooked.

The important differential diagnostic problems in relation to nonkeratinizing epidermoid carcinoma are produced by the more acute reactions of regeneration and repair within the respiratory tract. The cytology of these reactions has become more of a problem in recent years as bronchial brushing has grown more popular. Patients with a radiographic abnormality of the lung are likely to have this procedure. Direct sampling and good preservation of the cellular elements provide the cytopathologist with very active metaplastic and bronchial columnar epithelial cells having large nuclei and prominent nucleoli. Compare the two examples in Figures V-19 and V-20 with cases of nonkeratinizing epidermoid carcinoma. The cell arrangement in sheets is very similar. The nuclear chromatin structure is not as consistently distinct in every

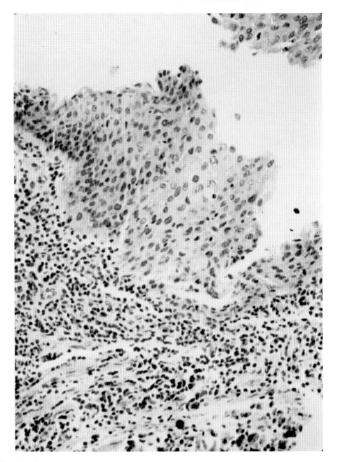

Fig. V-18. Histologic section of mycetoma cavity demonstrating atypical squamous metaplasia and probable source of the cells seen in Figure V-17. Hematoxylin and eosin stain, × 240.

cell, some cells in the benign atypias showing degenerative features. The cells, though active, have a reasonable nuclear cytoplasmic ratio, whereas this is greatly altered in the carcinoma cases (Figs. V-11 and V-12). Figure V-19 is from a case of pulmonary infarction, the cells originating from a reactive bronchial epithelium adjacent to the infarct. Figure V-20 is of a brushing of a small ulcer in a major bronchus. The etiology of the ulcer is undetermined.

The differential diagnostic problem of interpretation of cells having their source in regeneration and repair in the respiratory tract is not as troublesome with sputum and bronchial washings. Perhaps the greater degree of degeneration occurring with metaplastic and reactive cells in contrast to cancer cells, in these types of specimens, magnifies the differences in nuclear chromatin detail when the cells of the neoplasm are compared with atypical cells of repair. There are also quantitative differences, the regeneration and repair atypias exfoliating small numbers of cells in groups and the carcinomas yielding both single cells and cell groups in larger numbers.

It should be remembered that malignant squamous cells in a specimen of sputum

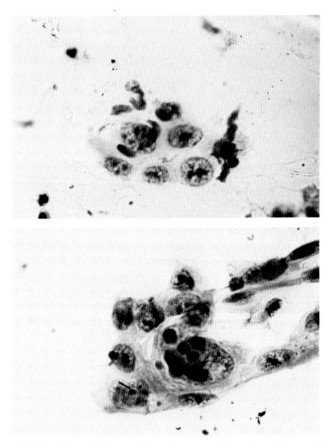

Fig. V-19. Bronchial brushing. Patient with pulmonary infarction. Sheets of cells with very prominent and angular nucleoli but pale somewhat irregular chromatin. Compare with brushings from nonkeratinizing epidermoid carcinoma (Figs. V-10 and V-11.) Note the differences in nuclear cytoplasmic ratio and variation in chromatin distinctness as seen in the case of infarction. Papanicolaou stain, × 600.

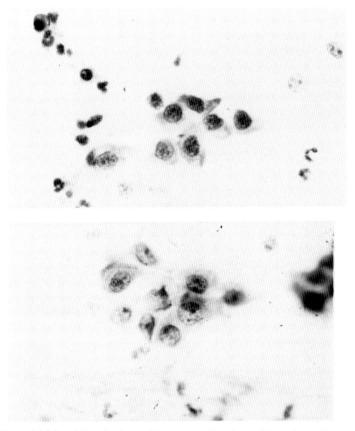

Fig. V-20. Bronchial brushing. Patient with nonspecific ulcer of bronchus. Sheets of cells with high nuclear cytoplasmic ratio but uniform chromatin. Compare nuclear detail and structure with brushings from nonkeratinizing epidermoid carcinoma. Papanicolaou stain, × 600.

need not come exclusively from neoplasms of the lung and bronchi. Carcinomas of the trachea, oral cavity, or even the pharynx and esophagus may be the source of malignant cells in sputum. A case of squamous-cell carcinoma of the esophagus is demonstrated in Figure V-21. There is marked degeneration of the single tumor cells, so that no cytoplasm is visible. The tumor cells are all smaller than those of typical keratinizing epidermoid carcinoma of the bronchus, but some of them have good nuclear detail with definite abnormal chromatin distribution. Note the extensive inflammatory background. That is the result of a bronchoesophageal fistula from extension of the esophageal carcinoma. Although there are convincing malignant features in these cells, their small size and the presence of the inflammatory background suggest that this sputum does not represent the usual cytology of a keratinizing epidermoid carcinoma of the bronchus. Coupling the cytologic features to clinical information that might suggest an esophageal neoplasm could lead to the correct cytologic diagnosis. This type of diagnostic approach is important for accurate pulmonary cytopathology.

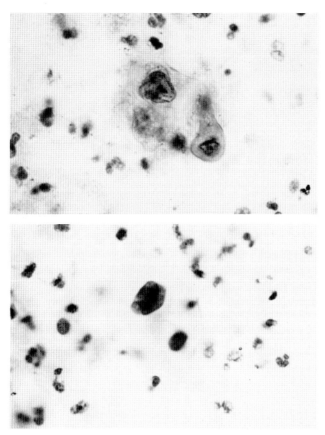

Fig. V-21. Sputum. Patient with keratinizing epidermoid carcinoma of the esophagus and a bronchoesophageal fistula from tumor invasion. Pleomorphic, well-preserved keratinizing tumor cells (upper panel), and degenerated naked nucleus from tumor cell (lower panel). Inflammatory background, relative paucity of tumor cells, and lack of very large pleomorphic keratinized cancer cells suggest that this sputum may *not* represent a primary lung cancer. Clinical correlation is needed to make a more specific diagnosis. Papanicolaou stain, × 600.

REFERENCES

1. U.S. Department of Health, Education, and Welfare: Public Health Service: *The Health Consequences of Smoking*, 1974. DHEW Publication No. (CDC) 74-8704. U.S. Superintendent of Documents, Washington, D.C., 1974.
2. Holbrook, J. H.: Tobacco and health. *CA, 27:* 344–353, 1977.
3. Larsson, S. and Zettergren, L.: Histological typing of lung cancer. Application of the World Health Organization classification to 479 cases. *Acta Pathol Microbiol Scand (A), 84:* 529–537, 1976.
4. Miller, A. B.: Recent trends in lung cancer mortality in Canada. *Can Med Assoc J, 116:* 28–30, 1977.
5. Doll, R. and Hill, A. B.: Mortality in relation to smoking: Ten year's observations of British doctors. *Br Med J, 5395:* 1399–1410, 1460–1467, 1964.
6. Doll, R. and Hill, A. B.: The age distribution of cancer: Implication for models of carcinogenesis. *J Royal Stat Soc, Series A, 134:* 133–166, 1971.
7. Whittemore, A. and Altshuler, B.: Lung cancer incidence in cigarette smokers: Further analysis of Doll and Hill's data for British physicians. *Biometrics, 32:* 805–816, 1976.
8. Berlin, N.: Summary and recommendations of the workshop on lung cancer. *Cancer, 33:* Suppl, 1744–1746, 1974.
9. Fontana, R. S., Sanderson, D. R., Woolner, L. B., et al.: The Mayo lung project for early detection and localization of bronchogenic carcinoma: A status report. *Chest, 67:* 511–522, 1975.
10. Beattie, E. J., Jr.: Lung cancer. *CA, 24:* 96–99, 1974.
11. Auerbach, O., Stout, A. P., Hammond, E. C., and Garfinkel, L.: Changes in bronchial epithelium in relation to cigarette smoking and in relation to lung cancer. *N Engl J Med, 265:* 253–267, 1961.
12. Cohen, M. H.: Guest Editorial. Lung Cancer: A status report. *J Natl Cancer Inst, 55:* 505–511, 1975.
13. Nellesheim, P.: Precursor lesions of bronchogenic carcinoma. *Cancer Res, 36*(PTZ)*:* 2654–2658, 1976.
14. Saccomanno, G., Archer, V. E., Auerbach, O., et al.: Development of carcinoma of the lung as reflected in exfoliated cells. *Cancer, 33:* 256–270, 1974.
15. Schreiber, H., Saccomanno, G., Martin, D. H., et al.: Sequential cytological changes during development of respiratory tract tumors induced in hamsters by Benzo (A) Pyrene-ferric oxide. *Cancer Res, 34:* 689–698, 1974.
16. Frable, W. J.: The relationship of pulmonary cytology to survival in lung cancer. *Acta Cytol, 12:* 52–56, 1968.
17. Kreyberg, L.: *Histological Typing of Lung Tumors.* International Histologic Classification of Tumors, No 1, World Health Organization, Geneva, 1967.
18. Nunez, V., Melamed, M.R., and Cahan, W.: Tracheo-bronchial cytology after laryngectomy for carcinoma of the larynx. II. Benign atypias. *Acta Cytol, 10:* 38–48, 1966.

CHAPTER

VI

Large-Cell Undifferentiated Carcinoma

Undifferentiated lung cancer composed of large anaplastic cells comprises approximately ten percent of the cases in most series.[1,2] Inclusion of cases in this group of tumors is probably a reflection of the diligence of the histologic sectioning of any given lung neoplasm in order to find differentiated areas of specific morphology.[3] From examination of the authors' own surgical pathology materials it is evident that there are cases of both squamous carcinoma and adenocarcinoma that have become so undifferentiated that morphologic features are no longer obvious enough to allow separation into one of the differentiated categories. The same problem is encountered with the cytopathology of these cases.

The cytologic diagnosis of these tumors as malignant is seldom difficult for any type of respiratory tract specimen. The cytologic criteria are listed in Table VI-1, and are those of obvious cancers. Perhaps in reviewing these cases there is too great a tendency to look for cytologic clues, such as prominence of nucleoli, depth of focus in cell groups, or density of cell cytoplasm and sharp cell borders that favor either adenocarcinoma or squamous-cell carcinoma. The cytologic criteria for a more specific diagnosis are usually being stretched in such cases. It is both accurate and reasonable, when confronted with large undifferentiated cells in respiratory specimens, simply to regard the lesion as a large-cell undifferentiated carcinoma.

An important feature of this type of tumor is the lack of consistency of the cytologic findings between different types of specimens from the same case. Figures VI-1 and VI-2 illustrate large undifferentiated malignant cells from the same tumor in sputum and bronchial washings. The tumor cells in sputum occur principally as single cells having a lobulated nucleus with an irregular nuclear membrane. A prominent nucleolus is usually seen and the chromatin is extremely irregular in distribution. These cells have a somewhat histiocytic appearance, but also have a very high nuclear cytoplasmic ratio that helps distinguish them from reactive alveolar pneumocytes. If the nucleus is degenerated, this distinction may be difficult (Fig. VI-3). The cytoplasmic borders are ragged and indistinct. Naked nuclei are frequently seen in sputum and cause problems in interpretation because of degeneration. The cytoplasm

175

Table VI-1
Cytologic Criteria: Large-Cell Undifferentiated Carcinoma

Large round or oval cells
Unequivocal malignant nuclear structure
Abundant single cells and cell clusters
Extensive tumor diathesis
Variation in cytoplasmic staining

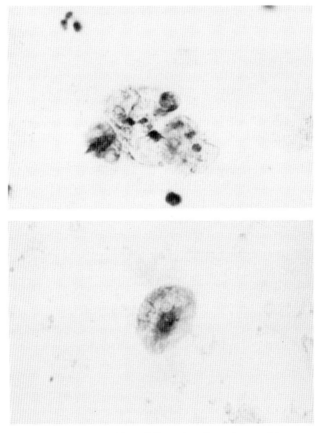

Fig. VI-1. Sputum. Cells from large-cell undifferentiated carcinoma. Note prominent nucleoli and pale but clumped chromatin. No cytoplasm is evident. One nucleus has a lobulated configuration. Papanicolaou stain, × 600.

Fig. VI-2. Bronchial washing. Same case as Figure VI-1. Tumor cells are more abundant and much more hyperchromatic. Irregular distribution of the chromatin is striking. Some finely vacuolated cytoplasm is visible. Papanicolaou stain, × 480.

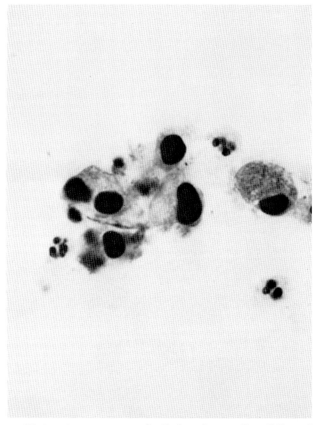

Fig. VI-3. Sputum. Histiocytic appearance of cells from large-cell undifferentiated carcinoma. Nuclei are hyperchromatic but without chromatin detail. Finely vacuolated cytoplasm accentuates the resemblance to reactive pneumonocytes. Papanicolaou stain, × 600.

is finely vacuolated or granular. The latter feature is seen more often in sputum which seems to accentuate the histiocytic appearance of the tumor cells (Fig. VI-3). Variation in staining quality of the cytoplasm, from basophilic to lightly eosinophilic, contributes to the similarity of the tumor cells to histiocytes. Many of the tinctorial changes as well as fragmentation and vacuolization of the cytoplasm are the result of degeneration. They are less apparent in bronchial washings and brushing specimens (Figs. VI-4, VI-5, and VI-6).

Brushing specimens provide large numbers of tumor cells, usually in sheets and/or loose clusters. Sometimes microbiopsies are seen (Fig. VI-6), with cells having a very high nuclear cytoplasmic ratio and a uniformly basophilic cytoplasm. Note the absence of sharp cell borders even in this well-preserved specimen. This is the feature most likely to separate this tumor from poorly differentiated epidermoid carcinoma (compare with Figs. V-11 and V-12).

DIFFERENTIAL DIAGNOSIS

The differential cytopathologic diagnosis in cases of large-cell undifferentiated carcinoma is most often one of correct classification of the tumor. Seldom are reactive

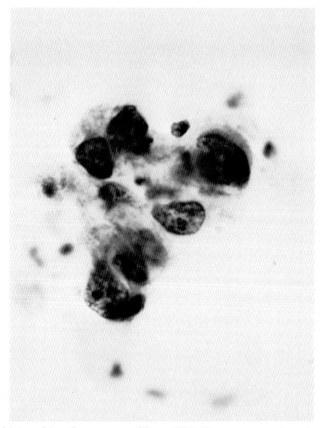

Fig. VI-4. Bronchial washing. Same case as Figure VI-3. Better preservation of the cells makes malignant characteristics obvious. Note absence of cell borders and very flimsy character of the cytoplasm. Papanicolaou stain, × 600.

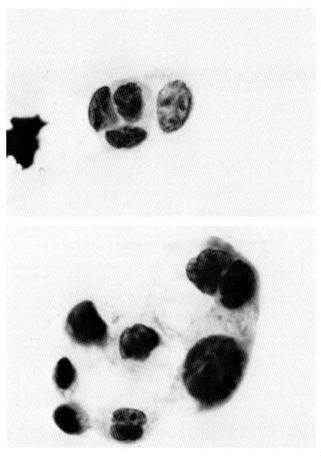

Fig. VI-5. Bronchial washing. Cells with better preserved cytoplasm and marked variation in cell size. No cell borders between adjacent cells are visible. Papanicolaou stain, × 600.

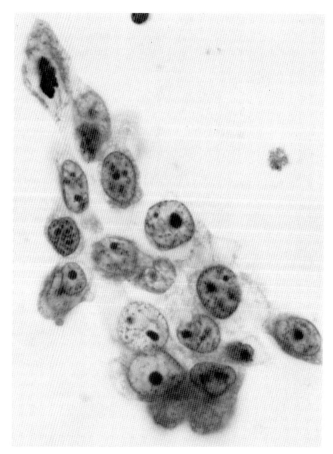

Fig. VI-6. Bronchial brushing. Good cell preservation and detail of large-cell undifferentiated carcinoma. Cells from what is essentially a microbiopsy. Papanicolaou stain, × 600.

lesions of the respiratory tract classified as this tumor type by mistake, since the malignant criteria of the cells in these cases are overwhelming. There are large numbers of tumor cells and unmistakable evidence of a tumor diathesis in the background of the smears. Misclassifications occur because tumor cells may cluster and have prominently lobulated nuclei, leading to a diagnosis of adenocarcinoma. If the tumor cells appear in broad sheets with some suggestion of sharp cell borders and the cytoplasm is both dense and uniform in consistency, poorly differentiated epidermoid carcinoma is suggested. These diagnostic errors are of no clinical importance, but they do lower the cytologic accuracy in reporting this tumor type to an undesirable level.

Several variations in presentation of this tumor are seen in Figures VI-7, VI-8, VI-9, VI-10, and VI-11. Figure VI-7 shows tumor cells in sputum exhibiting a histiocytic appearance, while the cells in Figure VI-9 are very pleomorphic. The nuclei in Figure VI-8 have degenerated to the extent that nucleoli are not visible. Despite the lack of nucleoli the cells still have other malignant features. Figures VI-9 and VI-10 reveal

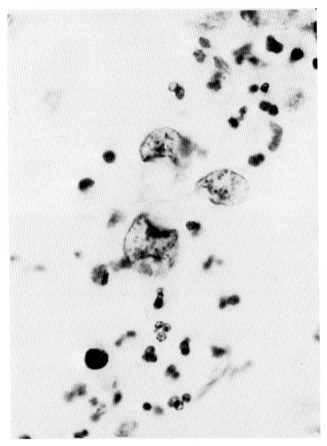

Fig. VI-7. Sputum. Large-cell undifferentiated carcinoma with single-cell histiocytic appearance in an inflammatory background. Irregular border of the nucleus and abnormal chromatin distribution are most significant features. Papanicolaou stain, × 600.

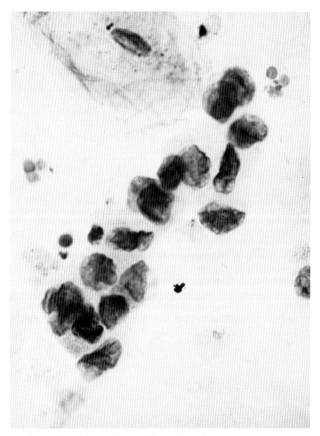

Fig. VI-8. Sputum. Syncytial clusters of undifferentiated tumor cells. Nuclei have degenerated to the extent that nucleoli are no longer visible. Papanicolaou stain, × 600.

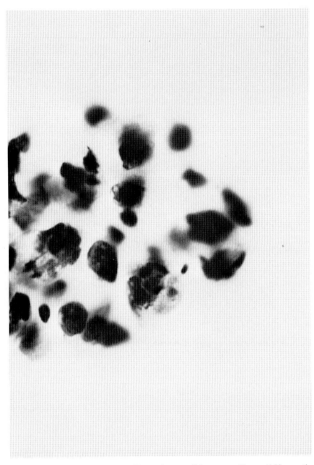

Fig. VI-9. Bronchial washing. Spindle-cell variant of large-cell undifferentiated carcinoma. Marked hyperchromasia and irregular, elongated configuration of nuclei are the diagnostic features. Papanicolaou stain, × 480.

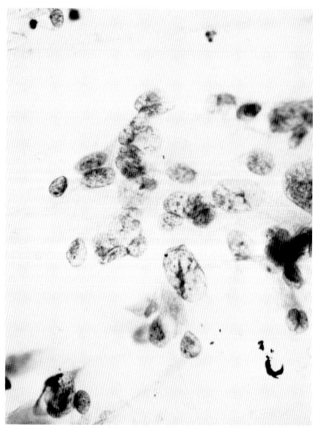

Fig. VI-10. Bronchial brushing. Same case as Fig. VI-9. Much better preservation of the cells retains the spindle-cell appearance with greater chromatin detail. Note very transparent cytoplasm. Papanicolaou stain, × 600.

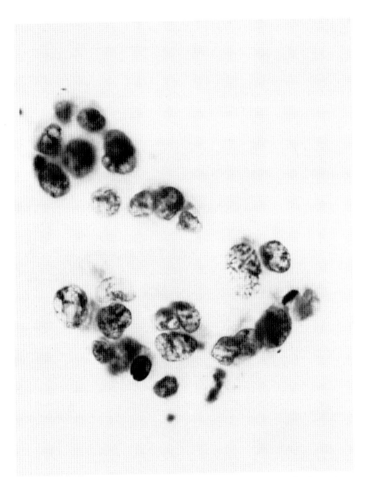

Fig. VI-11. Bronchial washing. Glandular configuration of tumor cells from large-cell undifferentiated carcinoma. Depth of focus is present. Cells have very high nuclear cytoplasmic ratio. Note multiple, but relatively small, nucleoli and many nuclei with some lobulation. Papanicolaou stain, × 480.

differences in cell preservation and detail from the same case of a large-cell undifferentiated carcinoma with a histologic pattern of spindle cells resembling sarcoma.
The bronchial washing shown in Figure VI-9 has many tumor cells, but the nuclei
exhibit a smudged chromatin pattern. Note the uneven nuclear border. The bronchial
brushing (Fig. VI-10) from the same case depicts both the variation in cell sizes and
the spindle-cell pattern to better advantage. Another unusual feature is the small size
of the nucleoli in these tumor cells. A clustering of cells with some depth of focus
and a glandular-like configuration may also be seen, as in Figure VI-11. It is not
unusual to diagnose these cases as poorly differentiated adenocarcinoma.

One further cytologic presentation of the large-cell undifferentiated carcinoma in
sputum relates to the neoplasm's propensity for rapid growth and extensive necrosis.
Figure VI-12 shows small irregular tumor cells in a sputum. There is little cytoplasm
associated with these contracted but pleomorphic tumor cells. There is granular
debris in the background of the smear. The pattern is very similar to a typical sputum
from oat-cell carcinoma (Chapter VII). There are a few cells in the depicted field,
however, that are beyond the size range of those seen with oat-cell carcinoma. This

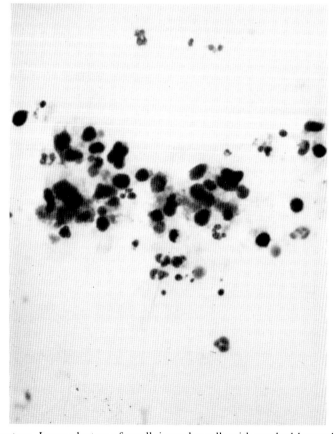

Fig. VI-12. Sputum. Loose clusters of small, irregular cells with marked hyperchromasia and
angular degenerated nuclei. Pattern in sputum of large-cell undifferentiated carcinoma simulating oat-cell carcinoma, the result of extensive tumor necrosis. Papanicolaou stain, × 600.

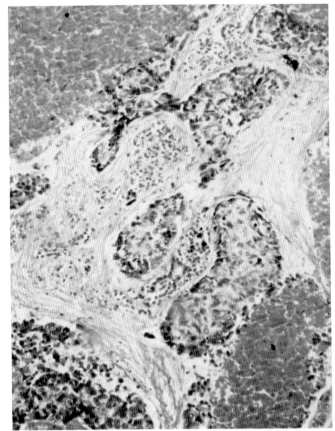

Fig. VI-13. Histology of tumor diagnosed from cells depicted in Figure VI-12. Note necrosis of tumor and the small, irregular pyknotic cells at the boundary between necrotic and viable tumor. Hematoxylin and eosin stain, × 250.

is the only clue to the correct diagnosis. The origin of these small cells is illustrated in Figure VI-13, representing marked necrosis in the tumor. This is a partially necrotic cancer with small degenerated cells at the boundary between the viable and dead tumor, its exfoliated cells giving rise to the false impression of oat-cell carcinoma.

REFERENCES

1. Auerbach, O., Garfinkel, L., and Parks, V. R.: Histologic typing of lung cancer in relation to smoking habits, year of diagnosis, and sites of metastases. *Chest, 67:* 382–387, 1975.
2. Hinson, K. F. W., Miller, A. B., and Tall, R.: An assessment of the World Health Organization classification of the histologic typing of lung tumors applied by biopsy and resected material. *Cancer, 35:* 399–405, 1975.
3. Churg, A.: The fine structure of large-cell undifferentiated carcinoma of the lung: Evidence for its relation to squamous-cell carcinoma and adenocarcinoma. *Human Pathol, 9:* 143–156, 1978.

Small-Cell Undifferentiated Carcinoma

CHAPTER

VII

Small-cell undifferentiated carcinoma (oat-cell carcinoma) is the most clinically aggressive lung cancer. This neoplasm accounts for about 15 to 20% of lung cancers.[1] The tumor originates in the primary and secondary bronchi or perihilar area of the lung in two-thirds of the cases.[2] The World Health Organization classification recognizes a polygonal and oval cell type, but this difference cannot as yet be appreciated in cytologic specimens.[3, 4] The prognosis is extremely poor for patients with small-cell carcinoma. Typically, they expire with widespread metastases.[2]

This neoplasm is thought to arise by malignant transformation of the reserve cells of the bronchial epithelium.[5-7] It is well known that some of the cells in the reserve cell layer have neurosecretory capabilities.[7, 8] Oat-cell carcinomas have been reported with hormonal manifestations documented clinically and biochemically.[2] Electron microscopic study of some of these tumors will demonstrate neurosecretory granules in the neoplastic cells, particularly those oat-cell carcinomas associated with ACTH production and Cushing's syndrome.[2, 7] It is important to recognize this tumor from cytologic specimens, since it is considered inappropriate to treat cases of small-cell carcinoma surgically, even if the lesion is clinically at an early stage. The tumor does respond to radiation therapy and there are some very preliminary data showing an increase in length of survival with combination chemotherapy.[2, 9] Long-term survivors of this tumor are both rare and anecdotal.

In classical cases, cytologic examination of sputum prepared by the Saccomanno technique will yield large numbers of small, very irregular tumor cells in loose aggregates. The background of the smear will usually be quite clean except for the immediate area around the tumor cells, where there will be a basophilic granularity composed of degenerated tumor cell cytoplasm. There may be a complete absence of histiocytes in these smears, but this does not mean that the tumor cells are contaminants. They should be in the same plane of focus as other cells seen in the smear.

The scant and fragmented cytoplasm leaves the nuclei either completely naked or with only a faint wisp of cytoplasm attached to one margin of the nucleus. Examination of the nuclei reveals that they are extremely hyperchromatic, with a markedly

189

irregular border. Perhaps because of the blending technique, the cells may have a blown-apart appearance, and some nuclei have large, visible blocks of chromatin. An occasional nucleolus may also be observed. Criteria for the diagnosis of small-cell undifferentiated carcinoma are summarized in Table VII-1.

Figures VII-1 and VII-2 illustrate the presentation of small-cell carcinoma in blended sputum described previously. In fresh sputum the presentation is somewhat different. There are usually fewer tumor cells and they occur in relatively compact groups. The irregularities of the nuclear border may not be as evident (Figs. VII-3, VII-4, and VII-5). Preservation of the cells may also be excellent in some cases (Figs. VII-3 and VII-4). The cell size is similar in both types of preparations (fresh and prefixed-blended), approximately 1½ times the diameter of a normal lymphocyte. Of major importance in fresh sputum is the nuclear molding of the tumor cells. The nuclei seem to push against each other, and the shape of the nucleus is distorted. Although this may also be seen in the prefixed and blended sputum, it is not a significant characteristic of that type of specimen (Fig. VII-2).

Bronchial washings usually show the same features cytologically as fresh sputum, regardless of whether they are prefixed and blended or handled as unfixed fresh specimens by smear and filter techniques. Nuclear molding is the most conspicuous feature. Variations among the cells in size and shape do occur, and these features should be carefully sought as additional evidence for malignancy in those cases where the nuclear molding and hyperchromasia are not convincing for a cancer diagnosis. Two cases diagnosed from bronchial washings are illustrated in Figures VII-6 and VII-7. Both show good nuclear molding. Figure VII-6 reveals greater variation in size and irregular shape of the tumor cells. Figure VII-7 depicts both nuclear molding and loose clusters of cells, but shows a remarkable uniformity of cell size. These variations do not seem to correlate with differences in histology of the tumors.

With bronchial brushing specimens it is necessary to readjust our concepts of the size of cells from small-cell undifferentiated carcinoma. The neoplastic cells appear to be two or three times the size of their counterparts as seen in sputum or bronchial washings (Fig. VII-8 and VII-9). This same phenomenon will also be seen in needle-aspiration biopsies of either lung or lymph nodes containing metastatic small-cell carcinoma. Compare Figure VII-10, a needle-aspiration biopsy of an oat-cell carcinoma in a supraclavicular lymph node, with Figure VII-7, the bronchial washing from the same case. Perhaps this difference in size is a result of excellent preservation of the cells or of sampling only viable tumor without degeneration and necrosis. The larger size of the nucleus provides more chromatin detail, which is seen to be quite coarse and granular. Several of the tumor cells demonstrate small nucleoli.

Table VII-1
Cytologic Criteria: Small-Cell Undifferentiated Carcinoma

Small cells in loose clusters
Dense hyperchromasia
Irregular nuclear border
Tumor diathesis only around tumor cells
Size variation of tumor cells with specimen type

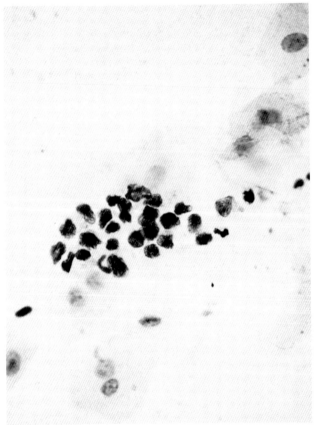

Fig. VII-1. Sputum. Cluster of small, undifferentiated malignant cells. Note variation in size and irregular nuclear border. Some chromatin clumping can be seen. Slight tumor diathesis is present around the malignant cells. No histiocytes were seen. Papanicolaou stain, × 600.

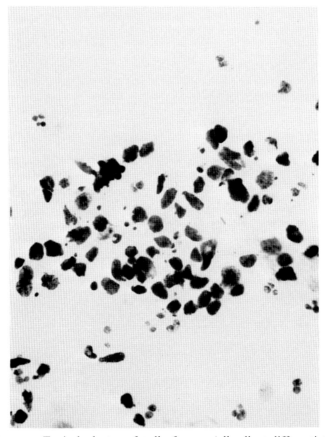

Fig. VII-2. Sputum. Typical cluster of cells from small-cell undifferentiated carcinoma, illustrating variation in shape and staining intensity. Only rarely has a cell any cytoplasm. Papanicolaou stain, × 600.

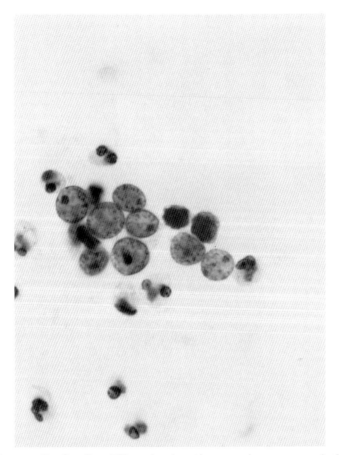

Fig. VII-3. Sputum. Small-cell undifferentiated carcinoma as it may appear in fresh sputum. Preservation in this particular example is excellent and even demonstrates several tumor cells with nucleoli. Nuclear molding is absent in this field. Papanicolaou stain, × 1000.

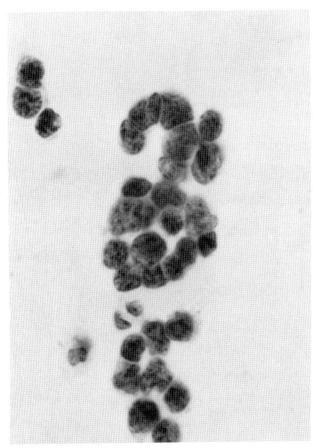

Fig. VII-4. Sputum. Fresh sputum from small-cell undifferentiated carcinoma in which the most conspicuous feature is nuclear molding. In several areas the cells are arranged in linear fashion (Indian file), with marked compression of the nuclei. Papanicolaou stain, × 1000.

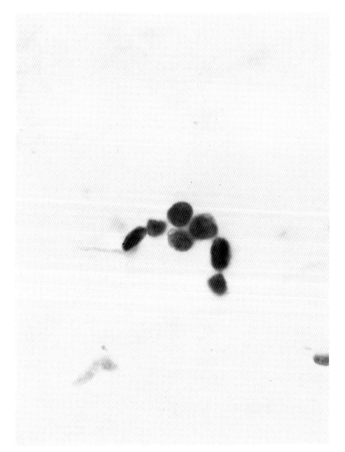

Fig. VII-5. Sputum. Fresh sputum from small-cell undifferentiated carcinoma. Shape of the cells is more typical of those cases from which the term *oat-cell carcinoma* is derived. Cells are also smaller than in other cases depicted as examples in fresh sputum and demonstrate much less nuclear detail. Papanicolaou stain, × 1000.

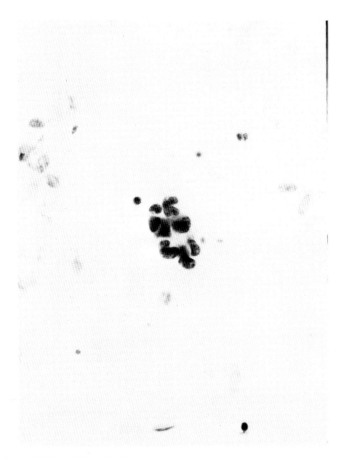

Fig. VII-6. Bronchial washing. Small, relatively tight cluster of small, undifferentiated malignant cells. Note nuclear molding and variation in chromatin distribution. Papanicolaou stain, × 600.

Fig. VII-7. Bronchial washing. Loose cluster of small, undifferentiated cells. All of the cells are intensely hyperchromatic. There is some nuclear molding, as well as variation in shape of the tumor cells. Papanicolaou stain, × 600.

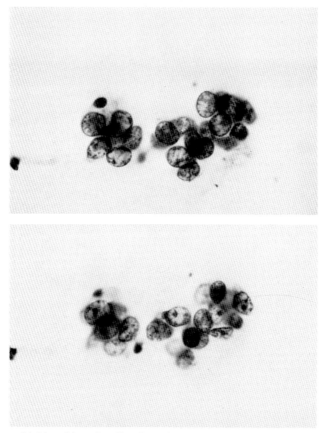

Fig. VII-8. Bronchial brushing. Small-cell undifferentiated carcinoma. Note the much larger size of the tumor cells and the excellent preservation, providing distinct nuclear chromatin. Some cells also have small nucleoli. Papanicolaou stain, × 600.

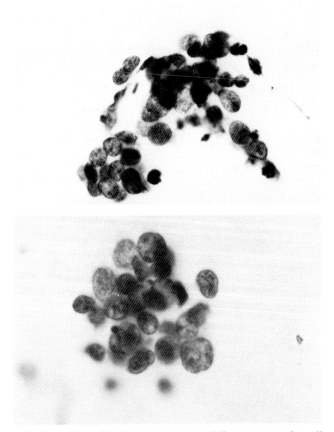

Fig. VII-9. Bronchial brushing. Specimens from two different cases of small-cell undifferentiated carcinoma. Cells are larger then the same cells seen in sputum and demonstrate good nuclear detail. There is both nuclear molding and some variation in focal plane of both cell clusters. Note the differences in nuclear size and shape. Papanicolaou stain, × 600 (upper panel) and × 1000 (lower panel).

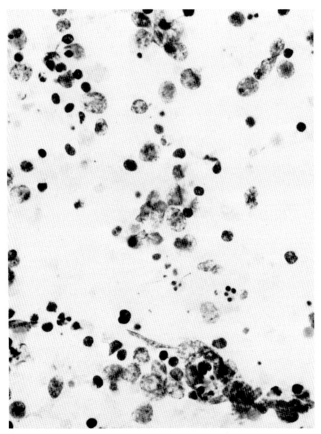

Fig. VII-10. Needle-aspiration biopsy. Supraclavicular lymph node with metastatic small-cell carcinoma. Note the same features of malignant cells as seen in brushing specimens. Bronchial washing, Figure VII-7, is from this patient. Note the difference in size of the tumor cells. Papanicolaou stain, × 600.

DIFFERENTIAL DIAGNOSIS

The differential diagnosis among cases of small-cell undifferentiated carcinoma most often involves reserve-cell hyperplasia. This reaction, previously described in Chapter III, occurs in conjunction with a variety of irritants of the respiratory tract. A comparative photograph (Fig. VII-11) reveals the cohesiveness and uniformity of the cells in reserve-cell hyperplasia contrasted to the dispersed cluster of cells from an oat-cell carcinoma. There is no basic difference in hyperchromasia. Both small-cell undifferentiated carcinoma and reserve-cell hyperplasia may be seen in a background of inflammation. This causes degeneration and loss of cohesion of all small cells, but has a more profound effect on the small, undifferentiated tumor cells. An example of oat-cell carcinoma from sputum in a background of inflammation is illustrated in Figure VII-12. Compare this case with the example of reserve-cell hyperplasia in Figure VII-11. For the case of small-cell carcinoma the tumor cells seem to blend in with the inflammatory cells, some of which have the appearance of

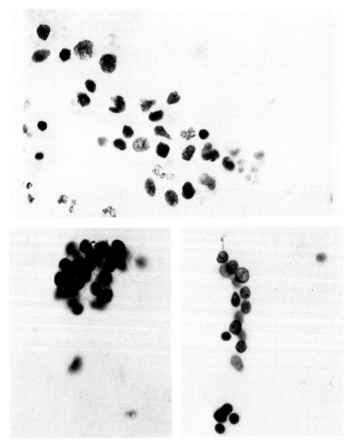

Fig. VII-11. Sputum. Comparison of cells from small-cell undifferentiated carcinoma (upper panel) and cells from reserve-cell hyperplasia (lower panels). Note variation in size, shape, and staining intensity of the tumor cells when compared with reserve cells. Papanicolaou stain, × 600.

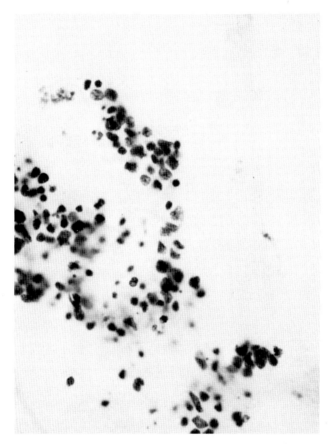

Fig. VII-12. Sputum. Small-cell undifferentiated carcinoma with an inflammatory background. Degeneration of the tumor cells is marked and they blend with the inflammatory cells, resembling small histiocytes. Papanicolaou stain, × 600.

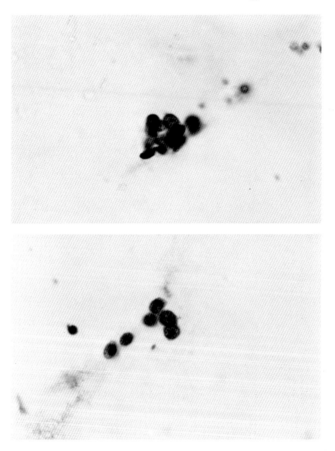

Fig. VII-13. Bronchial brushing. Cells from a bronchial carcinoid tumor. Cells are arranged in clusters, with some depth of focus. Note the presence of small nucleoli and granular chromatin of those nuclei in the plane of focus. Papanicolaou stain, × 600.

small histiocytes. Attention must be paid to the uneven nuclear border. Several clusters of tumor cells should be present with this smear pattern before a diagnosis of oat-cell carcinoma is rendered unequivocally.

Two other tumors composed of small cells may be encountered in pulmonary cytology and may be confused with oat-cell carcinoma, or else not considered neoplastic. One of the authors (Frable) has had experience with two cases of bronchial carcinoid tumor, one diagnosed as an atypical reserve-cell hyperplasia, the other as a carcinoid type of bronchial adenoma. Figures VII-13 and VII-14 illustrate these two cases. In Figure VII-13 the cells in both panels are cohesive, but have a three-dimensional configuration (depth of focus). Despite their cohesiveness, there is a visible chromatin structure. All of the cells, as they are brought into the focus, have a small nucleolus. The tumor cells therefore have some features of reserve cells and some of undifferentiated neoplastic cells. Similar cells are depicted in Figure VII-14, the second case, diagnosed correctly as a carcinoid type of bronchial adenoma. When coupled with the clinical findings of a small mass having an ulcerated surface

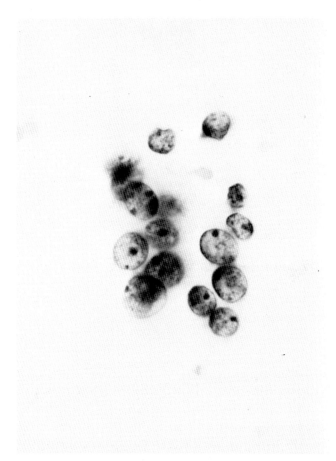

Fig. VII-14. Bronchial brushing. Cells from a bronchial carcinoid tumor correctly diagnosed. Cells are in a loose-cluster, almost acinar arrangement. They have a remarkable uniformity similar to reserve cells. Note the fine granularity of the nuclear chromatin and the presence of a small nucleolus in nearly every cell. Papanicolaou stain, × 1200.

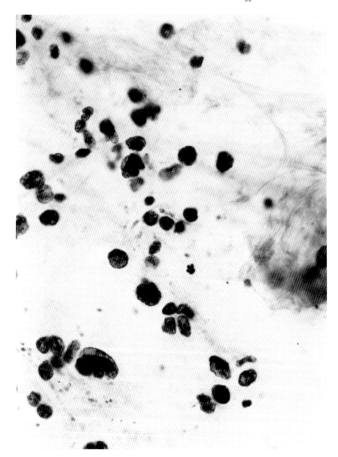

Fig. VII-15. Sputum. Cells from lymphosarcoma involving the lung diffusely. Note the similarity to small-cell carcinoma, including some nuclear molding. Cells are larger than most sputum cytology smears of oat-cell carcinoma. Papanicolaou stain, × 600.

projecting into the left lower lobe bronchus, the diagnosis should have been possible. The difficulty of this diagnosis has been described by others.[10]

Malignant lymphomas may involve the lung and exfoliate cells in the sputum. These cells may be small and pleomorphic, as seen in Figure VII-15, a case of poorly differentiated lymphosarcoma. Note the similarity to the sputum presentation of oat-cell carcinoma (Fig. VII-2). The malignant lymphocytes sometimes appear in the same area of the smear, thus suggesting that they are loosely cohesive. In the example demonstrated, there is even some suggestion of nuclear molding. From that pattern, the diagnosis of small-cell undifferentiated carcinoma is likely to be made. The correct diagnosis of lymphosarcoma depends both on the clinical findings and the presence of single cells with a morphology more in keeping with malignant lymphocytes rather than oat-cell carcinoma (Fig. VII-16). Obviously this diagnosis can be difficult.

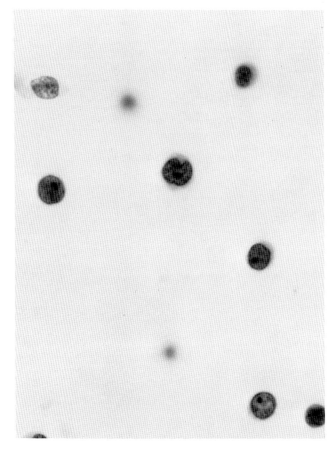

Fig. VII-16. Bronchial brushing. Lymphoblastic lymphosarcoma involving the lung. Even in the brushing specimen the cells occur as single cells usually characteristic of malignant lymphomas. One cell with an indented nucleus also suggests the lymphoma diagnosis. Papanicolaou stain, × 1000.

Reactive bronchial hyperplasia and the exfoliation of many *Creola bodies* (Chapter III) may occur in cases of small-cell undifferentiated carcinoma. The presence of clusters of hyperplastic bronchial cells may dominate the smear, catching the eye of the screener to the exclusion of recognition of small numbers of cells from oat-cell carcinoma. The cause of this hyperplastic response is unknown. When seen, particular care must be taken in examination of the smear to make sure that cells of small-cell undifferentiated carcinoma do not go undetected.

REFERENCES

1. Auerbach, O., Garfinkel, L., and Parks, V. R.: Histologic type of lung cancer in relation to smoking habits, year of diagnosis, and sites of metastases. *Chest, 67:* 382–387, 1975.
2. Eagan, R. T., Maurer, L. H., Forcier, R. J. et al: Small-cell carcinoma of the lung: Staging, paraneoplastic syndromes, treatment and survival. *Cancer, 33:* 527–532, 1974.

3. Hinson, K. F. W., Miller, A. B., and Tall, R.: An assessment of the World Health Organization classification of the histologic typing of lung tumors applied to biopsy and resected material. *Cancer, 35:* 399–405, 1975.

4. Kreyberg, L.: Histological Typing of Lung Tumours. In *International Histologic Classification of Tumors,* No. 1, World Health Organization, Geneva, 1967.

5. Ranchod, M.: The histogenesis and development of pulmonary tumorletts. *Cancer, 39:* 1135–1145, 1977.

6. Sayler, D. C., Sayler, W. R., and Eggleston, J. C.: Bronchial carcinoid tumors. *Cancer, 36:* 1522–1537, 1975.

7. Smith, L. H.: Oat-cell carcinoma as a malignant apudoma. *J Thorac Cardiovasc Surg, 70:* 147–151, 1975.

8. Weiss, L., and Creep, R. O.: *Histology.* 4th ed. McGraw-Hill Book Co., New York, 1977, pp 779–781.

9. Gilby, E. D., Bondy, P. K., Morgan R. L. et al: Combination chemotherapy for small-cell carcinoma of the lung. *Cancer, 39:* 1959–1966, 1977.

10. Kyriakos, M. and Rockoff, S. D.: Brush biopsy of bronchial carcinoid—A source of cytologic error. *Acta Cytol, 16:* 261–268, 1972.

VIII   Adenocarcinoma

In most series of primary lung cancer, adenocarcinomas account for about 15% of the cases.[1,2] There is some current evidence to suggest that the incidence of various types of lung cancer is changing.[3,4] Most remarkably, there is a very definite increase in the proportion of adenocarcinomas.[3] The recently noted increase in lung cancer in women has been accompanied by a parallel increase in adenocarcinomas, suggesting that sex may in part determine the neoplastic response of the lung to carcinogen, namely cigarette smoking.[4,5] We have been able to document both the increase in lung cancer in women and the increase in adenocarcinomas among our own cases.

Adenocarcinomas are usually found more peripherally in the lung. This depends in part on the histologic type and the theoretical cell of origin of this neoplasm.[6] Before the widespread use of bronchial brushing, the peripheral lesions were difficult to diagnose from conventional examination of bronchial washings and sputum. It is the experience of one of the authors (Frable) that, prior to the availability of bronchial brushing, adenocarcinoma diagnosed in bronchial washing or sputum cytology usually indicated an advanced lesion with lymph-node metastases. For those cases that were resected, it was frequently noted that, while the main neoplastic mass might be peripheral, the more proximal bronchi were usually encroached upon by metastatic nodes. One wonders if the malignant cells seen cytologically did not in some manner come from the area involved by lymph nodes about the more proximal bronchus rather than from the main tumor. Attempts to demonstrate actual communication or invasion of the bronchi by the tumor-bearing nodes are usually unsuccessful.

Adenocarcinomas may also present as pneumonic infiltrates or multiple nodules, the latter suggesting either granulomatous disease or metastatic tumor.[7] This combination of clinical presentations, coupled with the ability of the bronchial epithelium and alveolar lining cells to respond in dramatic fashion to a variety of pulmonary insults, makes both the diagnosis of adenocarcinoma of the lung and its separation from reactive conditions a challenging problem.

The authors believe that the various histologic types of adenocarcinoma can be either recognized or strongly suspected from the cytologic findings, admitting that some overlap in criteria will occur. Although this separation may be considered only an academic exercise, it provides some barriers against misdiagnosis of those proliferative and reactive processes in the lung that cytologically mimic adenocarcinoma.

TERMINAL BRONCHIOLO-ALVEOLAR CELL CARCINOMA

In an original study, one of the authors (Frable) found that typical pneumonic forms of terminal bronchiolo-alveolar cell carcinoma and even some single peripheral lesions exfoliate large numbers of cells both singly and in groups.[8] The cell groups averaged 16–18 cells per group. The cells were seen in a background of abundant histiocytes.[8] In some cases it was difficult to separate the tumor cells from the histiocytes. True histiocytic cells appearing in clumps are therefore particularly troublesome in the differential diagnosis. In some cases histiocytic-appearing cells in all probability represent degenerated tumor cells (Figure VIII-1).

In examples of well-differentiated terminal bronchiolo-alveolar cell carcinoma the

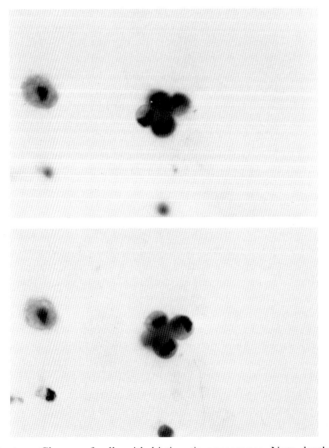

Fig. VIII-1. Sputum. Clumps of cells with histiocytic appearance. Note depth of focus and lack of distortion of the individual cells. These cells probably represent degenerated tumor cells from terminal bronchiolo-alveolar cell carcinoma. Papanicolaou stain, × 600.

tumor cells have a uniform appearance, with a finely granular nuclear chromatin and small round but distinct nucleoli. The nuclear cytoplasmic ratio is fairly high, but cytoplasm is easily seen. It is usually basophilic or amphophilic and finely vacuolated. There are no sharp cytoplasmic borders to the tumor cells, but neither do the cells form flat sheets of syncytial nature.

In an original communication, Frable emphasized the depth of focus required to see each tumor cell clearly within any group of neoplastic cells.[8] This appeared to be a fundamental characteristic of the cytology of terminal bronchiolo-alveolar cell carcinoma. While this is a remarkable feature, it may also be seen in some extreme reactions of bronchial cells and/or alveolar cells of the lung, for example following pulmonary infarction. The more important cytologic features of terminal bronchiolo-alveolar cell carcinoma are the excellent preservation of the tumor-cell morphology throughout the groups of cells and the retention of the individual and symmetrical shape of the neoplastic cells. There is no distortion or molding of the cells by neighboring cells. This is observed as one focuses through the cell group and sees the constant round or oval shape of each tumor cell (Figs. VIII-2 and VIII-3). The corresponding histiologic section for this case is illustrated in Figure VIII-4.

One characteristic of this tumor that varies with the type of cytologic preparation is vacuolization of the tumor cells. Sharply bordered vacuoles are very rare among the exfoliated cells of terminal bronchiolo-alveolar cell carcinoma if the specimen is prefixed and blended. Perhaps this is the effect of the alcohol or carbowax, since sharp vacuoles may be seen in the tissue sections from the same cases. If the specimen is prepared fresh, followed by fixation of the smear in 95% ethyl alcohol, tumor cells with sharply bordered vacuoles may be a major finding (Fig. VIII-5). This phenomenon has been documented as a major criterion of all types of primary adenocarcinoma of the lung.[9, 10] It is infrequently found if the specimen is prefixed.

Cells of terminal bronchiolo-alveolar cell carcinoma may appear to have cilia. These are not true cilia, since there is no terminal plate. Both transmission and scanning electron microscopy have revealed these tumor cells have many giant microvilli.[11] The microvilli provide the light microscopic effect of a fuzzy or ciliated border on some of the tumor cells (Figs. VIII-6 and VIII-7). This is more likely to be seen in bronchial brushing specimens, where preservation is often superior to either prefixed or fresh sputum or bronchial washings. A summary of the cytologic criteria for terminal bronchiolo-alveolar cell carcinoma is provided in Table VIII-1.

The World Health Organization classification has chosen the term *terminal bronchiolo-alveolar cell cell carcinoma*.[12] This is not to imply the sole cell of origin in the authors' opinion. A review of the recent literature would seem to substantiate that this tumor, which grows along the alveolar framework of the lung, may arise from ciliated cells of the terminal bronchiole, the nonciliated (Clara) cell of the terminal bronchiole, and, rarely, from the alveolar pneumocyte Type II.[13-17] Although the latter origin is the most controversial, it cannot be argued that sampling in all the cases studied by electron microscopy has always been in error.

More important to diagnostic cytology of lung cancer is the recent evidence suggesting that the growth pattern is not as important as the differentiation of the tumor cells.[18] Thus it is customary to think of terminal bronchiolo-alveolar cell carcinoma as a well-differentiated lesion presenting in lobar, single mass or multi-

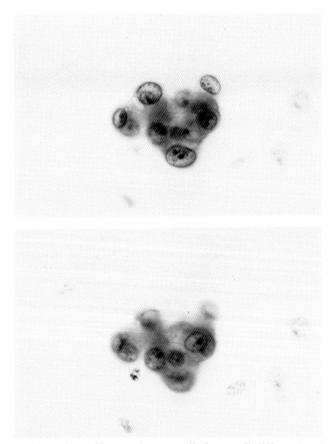

Fig. VIII-2. Bronchial washing. Cluster of tumor cells from well-differentiated terminal bron-chiolo-alveolar cell carcinoma photographed at different focal planes. Note good preservation of cell detail and lack of cell distortion. Same case as Figure VIII-1. Papanicolaou stain, × 600.

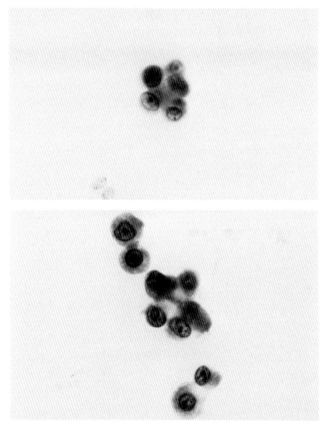

Fig. VIII-3. Bronchial brushing. Same cytologic characteristics as seen in bronchial washing from well-differentiated terminal bronchiolo-alveolar cell carcinoma. Same case as Figure VIII-2. Papanicolaou stain, × 600.

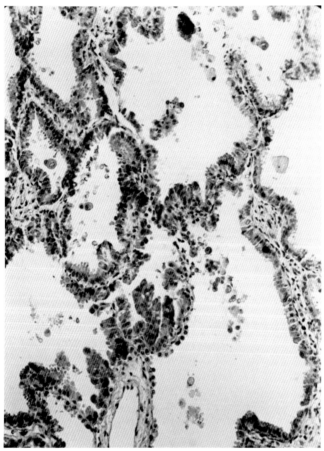

Fig. VIII-4. Section of well-differentiated terminal bronchiolo-alveolar cell carcinoma, the cells of which are illustrated in Figures VIII-1 through VIII-3. This lesion presented as a lobar infiltrate. Hematoxylin and eosin stain, × 200.

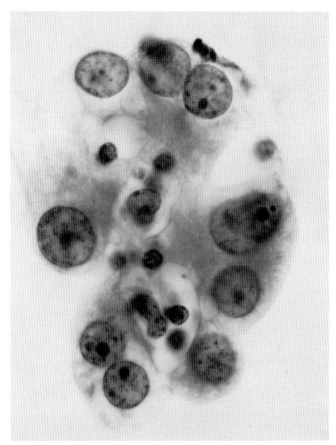

Fig. VIII-5. Sputum. Cluster of cells from terminal bronchiolo-alveolar cell carcinoma as seen in fresh sputum. Note presence of sharply bordered vacuoles. Papanicolaou stain, × 1000.

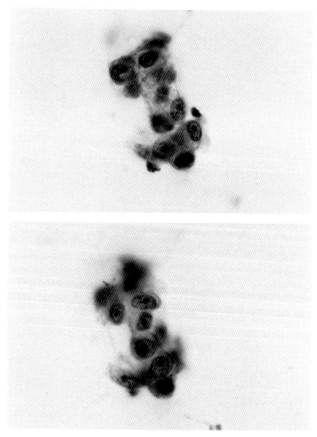

Fig. VIII-6. Bronchial washing. Two different focal planes of the same cell group from well-differentiated terminal bronchiolo-alveolar cell carcinoma. Note fuzzy cell border, giving the appearance of cilia. Papanicolaou stain, × 600.

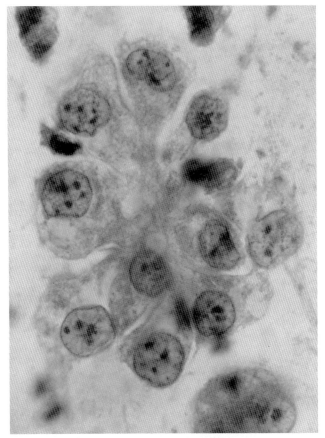

Fig. VIII-7. Sputum. Cells with finely vacuolated and irregular border having the appearance of cilia. Preparation of fresh sputum from a case of terminal bronchiolo-alveolar cell carcinoma. Papanicolaou stain, × 1000.

Table VIII-1
Cytologic Criteria: Adenocarcinoma—Terminal Bronchiolo-Alveolar Cell Carcinoma

Many cell clusters and single cells
Extreme depth of focus in cell clusters
Lack of cell and nuclear molding in clusters
Uniform symmetrical cells in clusters
Finely vacuolated cytoplasm
Distinct nuclear structure throughout
Uniform nucleoli

nodular form. However, an alveolar growth pattern may also be found for less differentiated lung carcinomas arising from cell types previously ascribed to the well-differentiated tumors.[18] Cytologically these poorly differentiated adenocarcinomas with alveolar growth patterns present many of the same criteria as the large-cell undifferentiated carcinomas.

Two cases of poorly differentiated adenocarcinoma of the lung with an alveolar growth pattern are depicted in Figures VIII-8 and VIII-10. Note the much larger size of the tumor cells. There is depth of focus present in these cell groups and the individual tumor cells are not molded. In contrast to the neoplastic cells in the well-differentiated terminal bronchiolo-alveolar cell carcinoma, these tumor cells have a somewhat lobulated nuclear outline. That seems to be a characteristic of the less differentiated forms of lung cancer with an alveolar growth pattern. From the tissue sections (Figs. VIII-9 and VIII-11) it is evident that many tumor cells are exfoliating, giving the appearance of a "carcinomatous pneumonia." Marked differences in prognosis have also been documented between this poorly differentiated alveolar type of carcinoma and its better differentiated counterpart.[18] This evidence also suggests that the growth pattern is not related to any particular cell of origin; differentiation is the most significant feature.

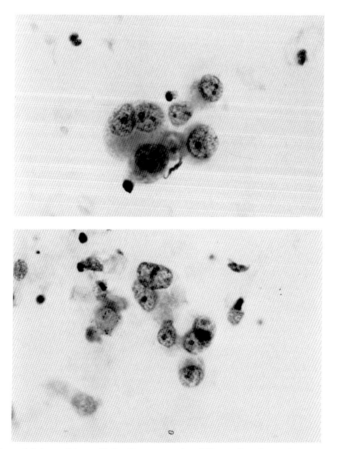

Fig. VIII-8. Bronchial washing. Cells from poorly differentiated carcinoma growing in an alveolar pattern. Compare with cells from well-differentiated cases depicted in Figures VIII-1, VIII-2, VIII-3, VIII-5, VIII-6, and VIII-7. Papanicolaou stain, × 600.

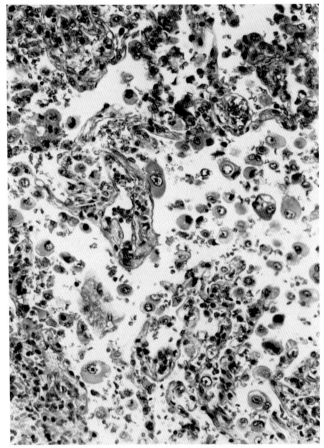

Fig. VIII-9. Histologic section of tumor diagnosed from cells illustrated in Figure VIII-8. Note alveolar growth pattern of this poorly differentiated carcinoma, with exfoliation of many cells into the alveolar lumen. Hematoxylin and eosin stain, × 225.

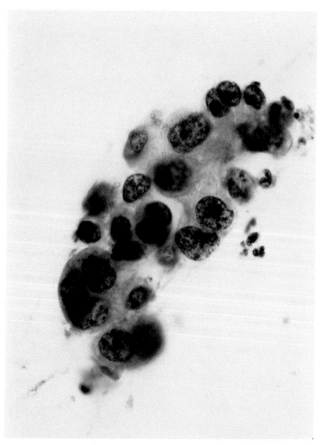

Fig. VIII-10. Bronchial washing. Cluster of poorly differentiated cells with evident depth of focus. Note lack of distortion of individual cells and lobulated border of the nuclei. Nucleoli are also more prominent than in well-differentiated terminal bronchiolo-alveolar cell carcinomas. Papanicolaou stain, × 600.

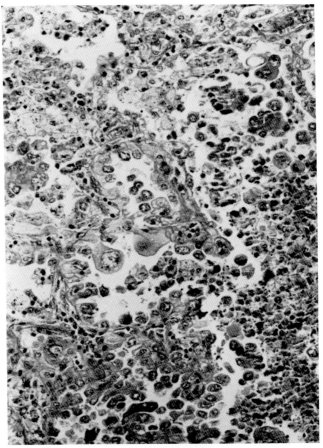

Fig. VIII-11. Histologic section of tumor diagnosed from cells in Figure VIII-10. This tumor is a very poorly differentiated carcinoma that is still growing along the alveolar framework of the lung. Hematoxylin and eosin stain, × 225.

BRONCHOGENIC PAPILLARY ADENOCARCINOMA

This rare form of bronchogenic carcinoma arises from the pseudostratified columnar epithelium of the bronchus, thus retaining some of that histologic form as a papillary neoplasm with columnar-shaped cells. The tumor has many similarities cytologically to terminal bronchiolo-alveolar cell carcinoma, except that it exfoliates fewer cells, but those cells have more cohesion within groups. The important cytologic criteria are summarized in Table VIII-2. The cell clusters have a definite but moderate three-dimensional characteristic, with insignificant cell distortion. Cell groups may exhibit some tendency to have a community border that is similar to cells of regeneration and repair in the tracheobronchial tree (compare Fig. VIII-12 with VIII-31). In well-preserved material, individual cells have definite borders. Nuclei are more hyperchromatic than those of well-differentiated terminal bronchiolo-alveolar cell carcinoma and exhibit larger nucleoli with less symmetry (Fig. VIII-14).

Table VIII-2
Cytologic Criteria: Adenocarcinoma—Bronchogenic Papillary

Both cell clusters and single cells
Moderate depth of focus of cell clusters
Usually uniform symmetrical cells
Relative lack of nuclear molding
Relative lack of community cell border
Clear or finely vacuolated cytoplasm
Distinct nuclear structure throughout

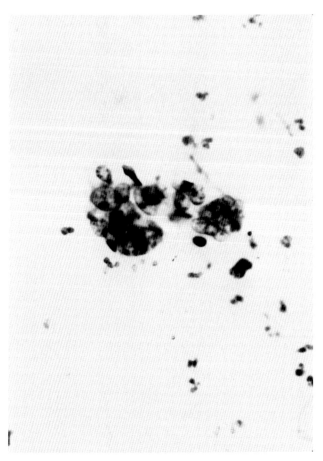

Fig. VIII-12. Bronchial washing. Cluster from bronchogenic papillary carcinoma. There is some community border among the cells. Chromatin is irregular and a few cells have large nucleoli. Preservation of the cells is less than optimal. Papanicolaou stain, × 600.

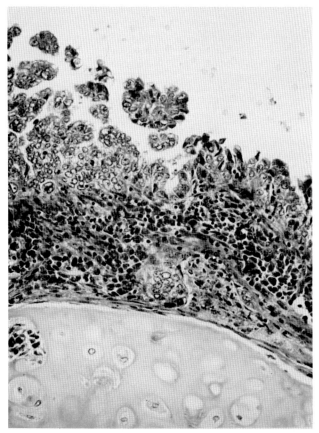

Fig. VIII-13. Histologic section of tumor diagnosed from cells in Figure VIII-12. A bronchogenic papillary carcinoma with intraluminal bronchial growth and invasion of the cartilage (lower left). Hematoxylin and eosin stain, × 240.

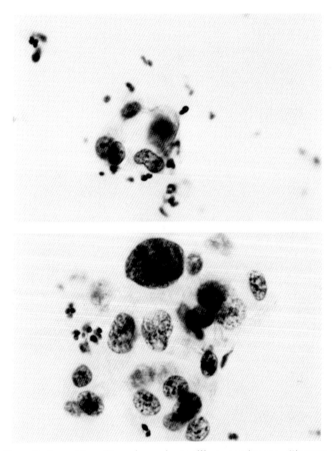

Fig. VIII-14. Bronchial washing. Bronchogenic papillary carcinoma. Clusters of tumor cells with variation in cell size, but with uniform shape. Clear or finely vacuolated cytoplasm is present. Papanicolaou stain, × 600.

The cytoplasm has a rather clear or finely vacuolated appearance, and, in prefixed specimens, there is an absence of sharply defined vacuoles. This is not true of fresh specimens, where many of the tumor cells may be expected to have cytoplasmic vacuoles. This vacuolated appearance of the neoplastic cells is quite evident in tissue section (Figs. VIII-13 and VIII-15). It should be emphasized that most important to the diagnosis of any of the lung adenocarcinomas is a consistency of abnormal nuclear structure and its preservation throughout all of the cells in any group. The degeneration so common to some of the cells of reactive bronchial hyperplasia and alveolar cell proliferations is not seen in well-preserved cell clusters from adenocarcinoma. The effects of degeneration and the difficulty in diagnosis can be seen in a cell group from a papillary bronchogenic carcinoma found in a sputum sample (Fig. VIII-16).

BRONCHOGENIC ACINAR ADENOCARCINOMA

Although there are typically far fewer tumor cells exfoliated by this neoplasm in comparison with terminal bronchiolo-alveolar cell carcinoma, malignant character-

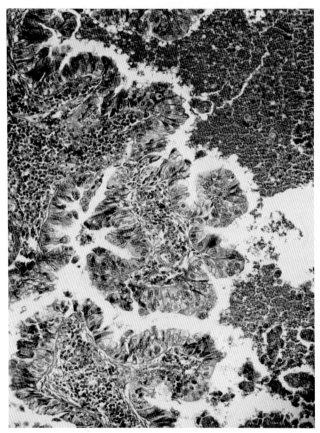

Fig. VIII-15. Histologic section of tumor diagnosed from cells in Figure VIII-14. The tumor has vacuolated cells, and some cells almost appear to have cilia. Hematoxylin and eosin stain, × 112.

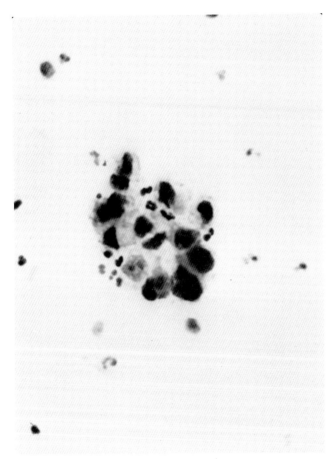

Fig. VIII-16. Sputum. Bronchogenic papillary carcinoma. Degeneration of the cells makes an exact diagnosis difficult. Variation in size and shape and irregular nuclear border suggest that these are malignant cells. Papanicolaou stain, × 600.

istics appear obvious. They are listed in Table VIII-3. The cytology of this type of lung cancer also varies with the degree of differentiation. The general characteristics of prominent nucleoli, high nuclear cytoplasmic ratio, and irregular clumping of chromatin are quite apparent. The tumor cells occur in sheets or groups, but in neither of these are there well-demarcated cell borders (Fig. VIII-18). The best description of these cells is that they form syncytia. Note also the loss of polarity of the individual nuclei within the syncytia (Figs. VIII-20, VIII-22, and VIII-23). These features serve to distinguish this tumor from poorly differentiated squamous-cell carcinoma, particularly as seen in bronchial brushings (Fig. VIII-25). Cytologically, that tumor has cells with definite borders and nuclei with pronounced orientation in the same direction. (Compare Fig. V-11).

If bronchogenic acinar carcinoma is well or moderately well differentiated, the cytoplasm of the tumor cells will stain either amphophilic or basophilic. This statement presumes good preservation of submitted specimens. If the tumor is poorly preserved, degeneration may cause the cells to become eosinophilic in sputum, while

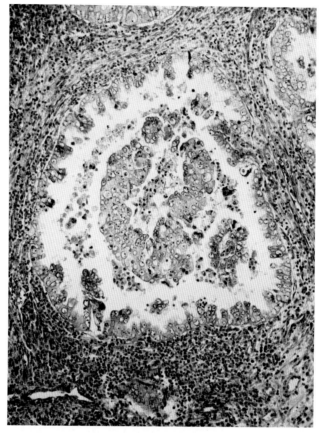

Fig. VIII-17. Histologic section of the tumor diagnosed from cells in Figure VIII-16. Broncho-genic papillary carcinoma. Hematoxylin and eosin stain, × 130.

Table VIII-3
Cytologic Criteria: Adenocarcinoma—Bronchogenic Acinar

Single and/or flat syncytia of cells
Lobulated nuclei
Macronucleoli and irregular nucleoli
Distinct nuclear chromatin
Variation in cytoplasmic staining according
 to degree of preservation
Finely vacuolated cytoplasm
Lack of sharp cell borders

they continue to stain basophilic in the usually better-preserved bronchial washings or brushings. Even if the cytoplasm assumes an eosinophilic appearance, it is finely vacuolated and does not have either the dense quality or deep pumpkin orange color of keratinizing squamous-cell carcinoma. The malignant cells seldom have definite borders, and the cytoplasm appears to fade at the periphery of the cell, actually fragmenting and disintegrating. Naked malignant nuclei are not unusual in the

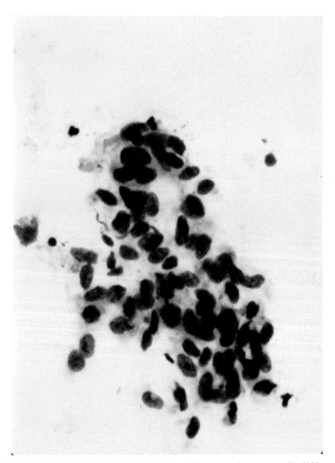

Fig. VIII-18. Bronchial washing. Bronchogenic acinar carcinoma well differentiated. Large irregular sheet of cells in syncytial arrangement. Note lack of well-defined cell borders and variation in polarity of the nuclei. Papanicolaou stain, × 600.

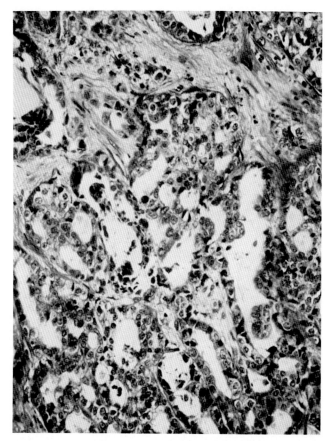

Fig. VIII-19. Histologic section of tumor diagnosed from cells in Figure VIII-18. Well-formed glands of adenocarcinoma arising from a major bronchus. Hematoxylin and eosin stain, × 240.

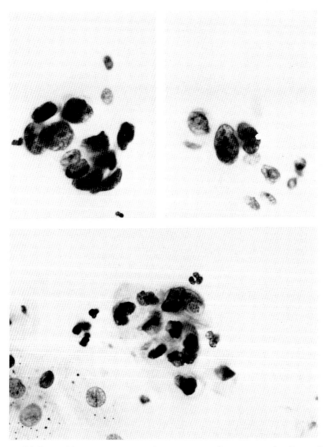

Fig. VIII-20. Bronchial washing (upper panels); sputum (lower panel). Bronchogenic acinar carcinoma, moderately well differentiated. The better preserved cells in sputum show syncytial arrangement and variation in shape. Cells have lost their cytoplasm in the bronchial washing, but still have prominent nucleoli and a very irregular nuclear border. Papanicolaou stain, × 600.

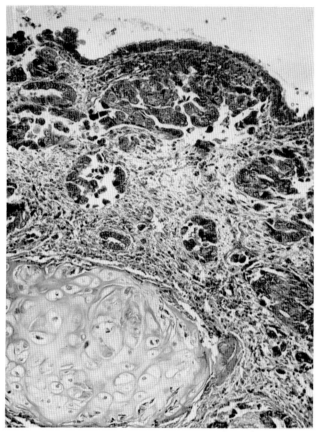

Fig. VIII-21. Histologic section of the tumor diagnosed from cells in Figure VIII-20. Moderately well-differentiated bronchogenic acinar carcinoma growing through and beneath attenuated bronchial mucosa and around cartilage. Hematoxylin and eosin stain, × 150.

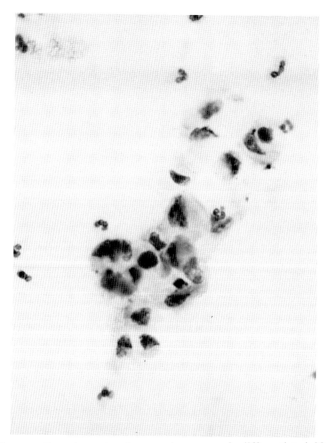

Fig. VIII-22. Sputum. Bronchogenic acinar carcinoma poorly differentiated. Nuclei are irregular and lobulated. The cytoplasm stains eosinophilic and is either finely vacuolated or poorly preserved. Papanicolaou stain, × 600.

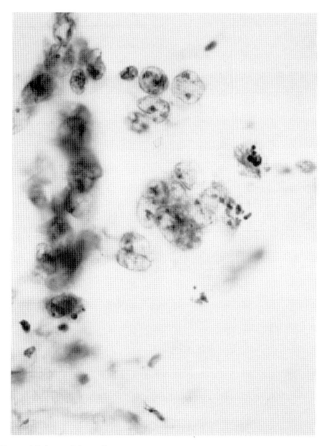

Fig. VIII-23. Bronchial washing. Same case as Figure VIII-22. Large, irregular cluster with more than the usual depth of focus for this type of adenocarcinoma. Nuclei are lobulated and have a pale but uneven and distinct chromatin pattern as seen in some conditions producing reactive alveolar pneumocytes or bronchial cells. Papanicolaou stain, × 600.

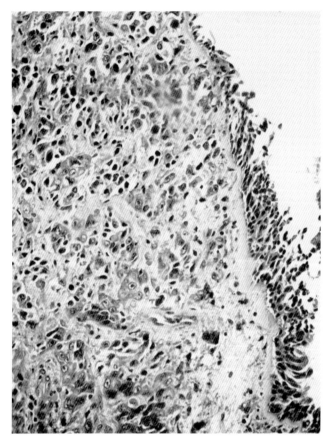

Fig. VIII-24. Histologic section of tumor diagnosed from cells in Figures VIII-22 and VIII-23. Poorly differentiated bronchogenic acinar carcinoma with vague glandular pattern seen beneath the bronchial mucosa. Hematoxylin and eosin stain, × 240.

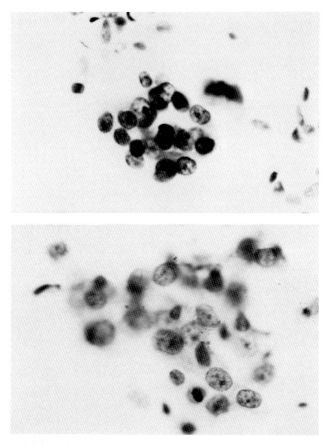

Fig. VIII-25. Bronchial brushing. Bronchogenic acinar carcinoma, poorly differentiated. Good preservation reveals cells with more uniform nuclei than seen in sputum and bronchial washings. Note absence of cell borders and relatively little cytoplasm, which stains basophilic. Nucleoli that are visible also vary in size. Papanicolaou stain, × 600.

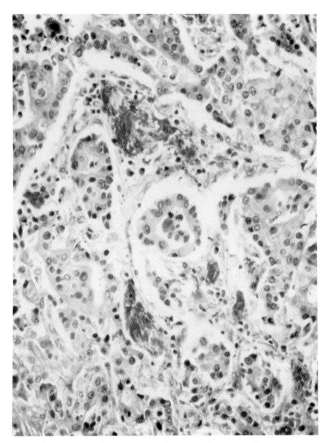

Fig. VIII-26. Histologic section of the tumor diagnosed from cells in Figure VIII-25. Poorly differentiated bronchogenic acinar carcinoma exhibiting one area where gland-like growth pattern is seen. Nuclear size, shape, and chromatin distribution correspond accurately to the brushing specimens. Hematoxylin and eosin stain, × 240.

cytology of this tumor but are rare in squamous-cell carcinomas. Figures VIII-18, VIII-20, and VIII-25 illustrate these features for different degrees of differentiation among examples of bronchogenic acinar adenocarcinoma.

Two examples of bronchogenic acinar carcinoma as seen in freshly collected sputum are demonstrated in Figures VIII-27 and VIII-28. The prominently vacuolated cytoplasm of most of the neoplastic cells is self-evident, emphasizing that consistent feature seen in adenocarcinomas without prefixation of the cytologic material. For the size of the cluster of cells in Figure VIII-27 there is relatively little depth of focus. The independent configuration of the neoplastic cells is well demonstrated in Figure VIII-28.

DIFFERENTIAL DIAGNOSIS

The capacity of the bronchial epithelium and alveolar pneumocytes to react to a host of stimuli has already been discussed. These reactive changes constitute the differential diagnostic problems encountered in cases of suspected adenocarcinoma of the lung. Some examples will illustrate the challenge in differentiating these reactions from carcinoma.

The classical *Creola Body* is the most commonly encountered papillary reactive group of bronchial epithelial cells to be confused with adenocarcinoma.[19] Bronchial hyperplasia been described and illustrated in detail elsewhere (see Chapter III). These groups of reactive cells have a good community border, extensive nuclear and cell molding, and at least some clearly defined cell borders (compare with Figs. VIII-12 and VIII-14, cases of papillary carcinoma). Close examination of most hyperplastic clusters of bronchial cells will reveal that some of them have either cilia or a terminal plate. It is also evident that the cells innermost within these clusters frequently lack good nuclear detail, in contrast to groups of malignant cells in papillary formations. The reactive hyperplastic cells are simply uniformly hyperchromatic, accentuating the overall staining of the cluster. This feature is noted first in screening but is the least reliable for a diagnosis of malignancy.

Several other specific reactions may produce either hyperplastic clusters of bronchial cells or reactive alveolar pneumocytes.[20] These are virus pneumonias, and endotracheal aspiration in patients with pneumonia or congestive heart failure,

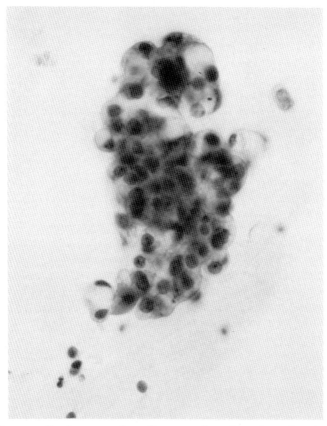

Fig. VIII-27. Sputum. Fresh preparation of sputum from bronchogenic acinar carcinoma, demonstrating vacuolated tumor cells. There is relatively little depth of focus for the size of the cell cluster. Papanicolaou stain, × 400.

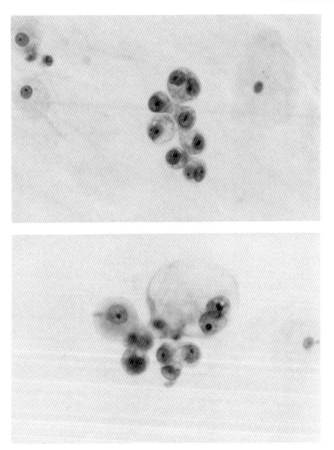

Fig. VIII-28. Sputum. Fresh preparation of sputum from another case of bronchogenic acinar carcinoma. Both cell groups demonstrate a few sharply bordered vacuoles. The cells also exhibit an independent configuration, though there is no depth of focus. Cell arrangement is somewhat suggestive of terminal bronchiolo-alveolar cell carcinoma. Papanicolaou stain, × 400.

pulmonary infarction, and occasionally with granulomatous disease of the lung. Examples of reactive cells from an adenovirus infection are illustrated in Figure VIII-29. The cell cluster in the upper panel of the figure was obtained from a bronchial brushing in a patient with a hazy infiltrate in both lungs, fever, and dry cough. The granularity of the chromatin in these nuclei is pronounced and is coupled with some depth of focus. Small nucleoli are noted, standing out in stark contrast to the relatively pale chromatin. The bottom half of the figure depicts cells from sputum in this case. Note the contrast in relation to the cells from the bronchial brushing. The additional feature of numerous inflammatory cells, some of which are within the cytoplasm of the reactive cells, is an important differential finding, indicating a reactive process. Figure VIII-30 demonstrates similar reactive cells, in a large sheet, also from this case. This type of exfoliation is common in bronchial washing specimens from adenovirus infections. The difference between the pale but distinct chromatin and small nucleoli is evident. Though the cytoplasmic borders cannot be

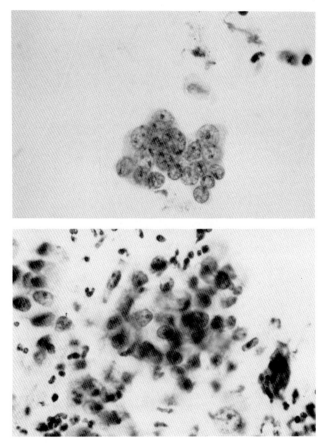

Fig. VIII-29. Bronchial brushing (upper panel) and sputum (lower panel). Reactive cells from a patient with an adenovirus infection. Note marked variation in appearance of the cells between the two different types of specimens. Compare the cells in the upper panel to cases of terminal bronchiolo-alveolar cell carcinoma (Figs. VIII-2, VIII-3, VIII-6, and VIII-8). Papanicolaou stain, × 600.

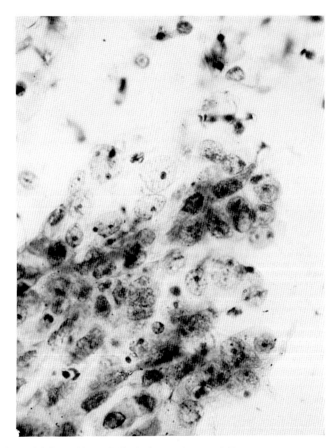

Fig. VIII-30. Bronchial washing. Cells in large sheet exfoliated from the same patient as in Figure VIII-29. Adenovirus infection of the lung. Note the prominence of the nucleoli within a very pale chromatin background to the nucleus. Scattered inflammatory cells are seen in the cytoplasm of these reactive cells. Papanicolaou stain, × 600.

seen, there is usually adequate cytoplasm for reactive cells, the nuclear cytoplasmic ratio not being significantly altered. The contrasting pattern of the cytology among the three types of specimens is quite helpful in arriving at a benign diagnosis. This inconsistency of cytomorphology is not as apparent in cases of adenocarcinoma. The very broad sheets of cells, showing pale chromatin with contrasting prominent nucleoli, should in themselves lead to the consideration of an adenovirus infection.

Sputum, bronchial washings, and/or brushings obtained from patients with recent pulmonary infarction provide the most difficult reactive cells to interpret.[21] Figure VIII-31 depicts cells from such a case. Note the reactivity of the nuclei and the granularity of the chromatin. There is, however, an inconsistency in the cells from the two groups in Figure VIII-31, with some having very smudged nuclei and others having large vacuoles, while some of the cells exhibit nuclear molding and excellent cohesion within the cluster. All of these features, when compared with cells from adenocarcinoma, should be helpful in recognizing reactive cells in cases of pulmonary infarction. In this particular patient the cells disappeared and the chest film showed

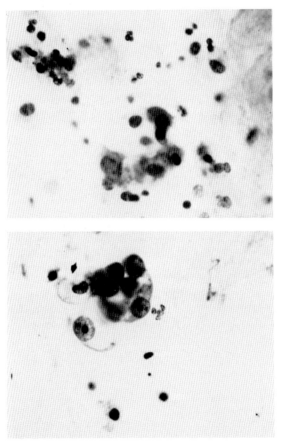

Fig. VIII-31. Sputum. Reactive alveolar pneumocytes from a patient with clinical pulmonary infarction. Note the sharply bordered vacuoles in the cytoplasm and the variation in chromasia and chromatin detail among the cells. Papanicolaou stain, × 600.

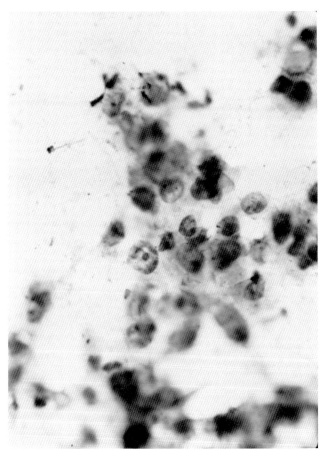

Fig. VIII-32. Bronchial washing. Cells exfoliated from patient with multiple pulmonary emboli and microscopic infarcts. Clinical picture and radiographs suggested lymphangitic spread of tumor. Note variation in preservation and nuclear detail of these cells. There is a definite resemblance to cells of poorly differentiated bronchogenic acinar adenocarcinoma. Papanicolaou stain, × 600.

resolution of a large lung mass about three weeks after the initial symptoms of hemoptysis. The patient at that time had evidence of peripheral thrombophlebitis.

Another example of pulmonary infarction presented a dramatic radiograph with a heavy infiltrative process that could easily be interpreted as lymphangitic spread of carcinoma. The patient exfoliated large numbers of pleomorphic cells, primarily in sheets but exhibiting both cell and nuclear molding. These cells were extremely hyperchromatic (Fig. VIII-32), but lacked internal nuclear detail. Some had prominent nucleoli and others did not. This variation of the cytomorphology was the diagnostic feature that suggested only reactive cells. Tissue obtained from open-lung biopsy showed multiple small pulmonary emboli and extremely reactive alveolar cells adjacent to microscopic infarcts (Figure VIII-33).

Bronchial brushings have not been of help in this type of case.[21] In fact the cells appear even more reactive, perhaps because of better preservation and the presence

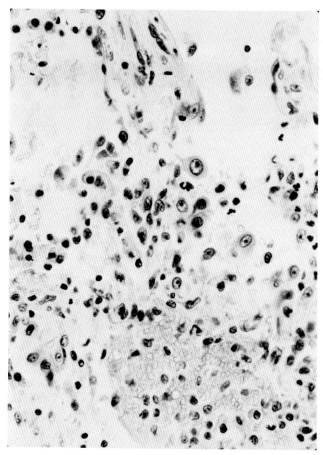

Fig. VIII-33. Tissue from lung biopsy, illustrating reactive alveolar pneumocytes next to microscopic infarcts. These cells are probably the source of those illustrated in Figure VIII-32. Hematoxylin and eosin stain, × 375.

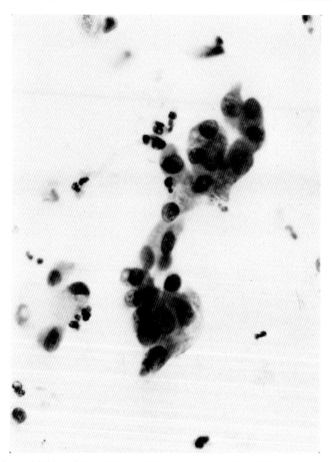

Fig. VIII-34. Bronchial brushing. Patient with pulmonary infarct. Very reactive cells are either alveolar pneumocytes or bronchial cells. Note good cohesion of the cells and some cells with cell and nuclear molding. Papanicolaou stain, × 600.

of many more of them obtained by the brushing technique. Figure VIII-34 illustrates one example. The cells are markedly hyperchromatic and the nuclear cytoplasmic ratio is high. The chromatin is *not* uniformly granular, but appears smudged in some of the cells. Good nuclear and cell molding can be seen, and there is only an insignificant three-dimensional configuration to the cells. Like examples of cells from infarcts of the lung in sputum and bronchial washings, brushings will also reveal inconsistent abnormalities.

Figures VIII-35, VIII-36, and VIII-37 represent additional examples of reactive alveolar pneumocytes from a case of blastomycosis, desquamative interstitial pneumonitis, and tracheal aspiration in a patient with congestive heart failure. The general cytologic features of these cell sheets and clusters suggest regeneration and tissue repair. They are all relatively active-looking cells, but have an essentially normal nuclear cytoplasmic ratio. Noteworthy are the prominent nucleoli, which are smooth and round, except for the case of blastomycosis. In that case the chromatin is pale, whereas in the other examples the nuclear chromatin is dense and smudged. There

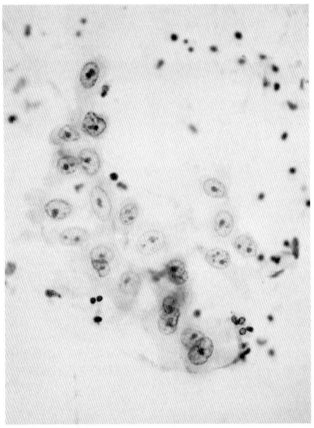

Fig. VIII-35. Sputum. This patient exfoliated very reactive cells suggestive of tissue repair. Note the prominent nucleoli within an extremely pale chromatin background of the nucleus. The patient had widespread blastomycosis of the lung. Papanicolaou stain, × 600.

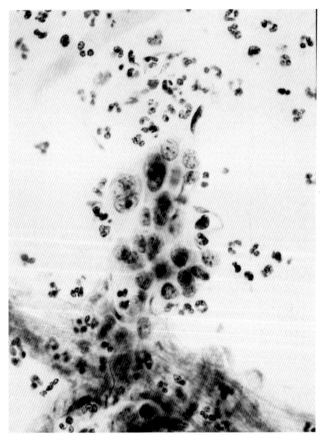

Fig. VIII-36. Tracheal aspiration. Aspiration of reactive alveolar pneumocytes from a patient in chronic congestive heart failure requiring a tracheostomy. There are large numbers of polymorphonuclear leukocytes, some within the cytoplasm of the reactive cells. Papanicolaou stain, × 600.

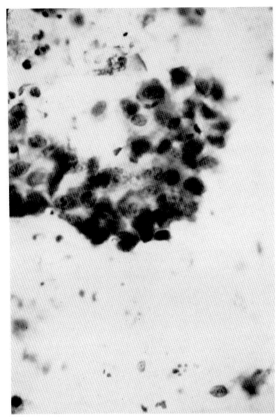

Fig. VIII-37. Bronchial washing. Large cluster of reactive alveolar pneumocytes exfoliated from this patient with desquamative interstitial pneumonitis. Note the variation in chromatin detail and the degenerated appearance of many of the nuclei. Papanicolaou stain, × 600.

is little depth of focus for those cells found in clusters. Polymorphonuclear leukocytes are seen within the cytoplasm of the cells or in large numbers in the surrounding background of the smear. In summary, the basic features of cells seen in regeneration and repair, regardless of the type of cytologic specimen, should be kept in mind to avoid an incorrect diagnosis of carcinoma in respiratory cytopathology.

REFERENCES

1. Auerbach, O., Garfinkel, L., and Parks, V. R.: Histologic type of lung cancer in relation to smoking habits, year of diagnosis and sites of metastases. *Chest, 67*(4): 382–387, 1975.
2. Hinson, K. F. W., Miller, A. B., and Tall. R.: An assessment of the World Health Organization classification of the histologic typing of lung tumors applied to biopsy and resected material. *Cancer, 35:* 399–405, 1975.
3. Holbrook, J. H.: Tobacco and health. *CA 27:* 344–353, 1977.
4. Vincent, R. G., Pickren, J. W., Lane, W. W., Bross, I., Takita, H., Horten, L., Gutierrez, A. C., and Pzepka, T.: The changing histopathology of lung cancer. A review of 1,682 cases. *Cancer 39:* 1647–1655, 1977.
5. Miller, A. B.: Recent trends in lung cancer mortality in Canada. *Can Med Assoc J 116*(1): 28–30, 1977.
6. Feinstein, A. R., Geffman, N. A., and Yesner, R.: The diverse effects of histopathology on manifestations and outcome of lung cancer. *Chest 66:* 225–229, 1974.

7. Donaldson, J. C., Kaminsky, D. B., and Elliott, R. C.: Bronchiolar carcinoma. Report of 11 cases and review of the literature. *Cancer 41:* 250–258, 1978.
8. Smith, J. H. and Frable, W. J.: Adenocarcinoma of the lung. Cytologic correlation with histologic type. *Acta Cytol 18:* 316–320, 1974.
9. Foot, N. C.: Identification of types of pulmonary cancer in cytologic smears. *Am J Pathol 28:* 963–983, 1952.
10. Spjut, H. J., Frier, D. J., and Ackerman, L. V.: Exfoliative cytology and pulmonary cancer. A histopathologic and cytologic correlation. *J Thor Surg, 30:* 90–107, 1955.
11. Johnston, W. W., Ginn, F. L., and Amatulli, J. M.: Light and electron microscopic observations on malignant cells in cerebrospinal fluid from metastatic alveolar cell carcinoma. *Acta Cytol 15:* 365–371, 1971.
12. Kreyberg, L.: Histological Typing of Lung Tumors. In *International Histologic Classification of Tumours*, No. 1, World Health Organization, Geneva, 1967.
13. Bedrossian, C., Weilbaecher, W. M., Bertinck, D. G., and Greenberg, S. D.: Ultrastructure of human bronchiolo-alveolar cell carcinoma. *Cancer 36:* 1399–1413, 1975.
14. Greenberg, S. D., Smith, M. N., and Spjut, H. J.: Bronchiolo-alveolar carcinoma. Cell of origin. *Am J Clin Pathol 63:* 153–167, 1975.
15. Jacques, J. and Currie, W.: Bronchiolo-alveolar carcinoma: A Clara cell tumor. *Cancer 40:* 2171–2180, 1977.
16. Lefer, L. G. and Johnston, W. W.: Electron microscopic observations on sputum in alveolar cell carcinoma. *Acta Cytol 20:* 26–31, 1976.
17. Torikata, C. and Ishiwata, K.: Intranuclear tubular structures observed in cells of an alveolar cell carcinoma of the lung. *Cancer 40:* 1194–1201, 1977.
18. Tao, L. C., Delarue, H. C., Sanders, D. and Weisbrod, G.: Bronchiolo-alveolar carcinoma. A correlative clinical and cytologic study. *Cancer 42:* 2759–2767, 1978.
19. Naylor, B. and Railey, C.: A pitfall in the cytodiagnosis of sputum of asthmatics. *J Clin Pathol 17:* 84–89, 1964.
20. Berkheiser, S. W.: Bronchiolar proliferation and metaplasia associated with bronchiectasis, pulmonary infarcts and anthracosis. *Cancer 12:* 499–508, 1959.
21. Scoggins, W. G., Smith, R. H., Frable, W. J., and O'Donohue, Jr., W. J.: False-positive cytological diagnosis of lung carcinoma in patients with pulmonary infarcts. *Ann Thorac Surg 24:* 474–480, 1977.

CHAPTER

IX

Other Neoplasms of the Lung, Primary and Metastatic

The cytopathologist must be concerned with other neoplasms of the lung in addition to the major cancers just described. These may be grouped into three broad categories: unusual primary neoplasms of the lung, both carcinomatous and sarcomatous; neoplasms involving the respiratory tract by direct extension or metastases; and lymphomas that are usually secondary tumors, but may occasionally be primary in the lung. Strictly within the cytopathologist's domain must be included malignant tumors that are primary in structures contiguous to the respiratory tract, the oral cavity, and, more remotely, the esophagus. Neoplasms in these areas may contaminate spontaneously produced sputum with malignant cells. Although none of the tumors to be described are common in terms of cytopathology, they must be considered as part of the differential diagnosis in any specimen with abnormal cells from the respiratory tract.

UNUSUAL PRIMARY CARCINOMAS AND SARCOMAS OF THE LUNG

The most common cancer of this relatively rare group of neoplasms is the mixed carcinoma composed of glandular and squamous epithelium, also referred to as mucoepidermoid carcinoma. This malignancy is probably diagnosed with a frequency that is in proportion to the diligence of sampling of any lung tumor or to the number of special stains performed to determine the presence of mucin positive cells. Thus the incidence in any series of lung cancers may range from less than 1% to 1.5%.[1,2]

Cytologically, mixed adenosquamous-cell carcinoma presents with a pattern of cells suggesting adenocarcinoma of bronchogenic acinar type. Several examples of cells from cases of this type of carcinoma are illustrated in Figures IX-1 through IX-3. The epidermoid features of the depicted cells are scanty and likely to be overlooked. An exception is noted in Figure IX-2. The tumor cells occur in small sheets or as single cells. Within the sheets there may be some cells with a prominence of the cytoplasmic border and/or sharp margins of cytoplasm between adjacent cells (Fig. IX-1, arrow). The cytoplasm of the epidermoid carcinoma cells has a dense quality, while other cells in the sheet have a finely vacuolated cytoplasm. It is rare to see any

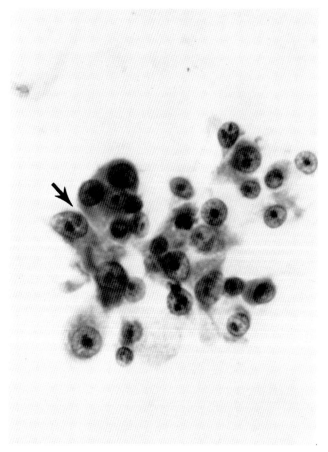

Fig. IX-1. Bronchial brushing. Cells from mucoepidermoid carcinoma. Note cells with finely vacuolated cytoplasm adjacent to cells with dense cytoplasm and sharp cytoplasmic borders (arrow). Papanicolaou stain, × 600.

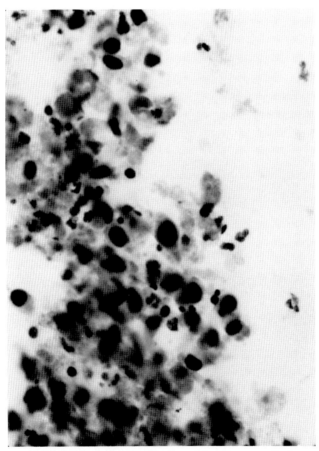

Fig. IX-2. Sputum. Single field of pleomorphic keratinized tumor cells from a case of muco-epidermoid carcinoma. Most of these cases contain only a few easily identifiable squamous cancer cells. Papanicolaou stain, × 600.

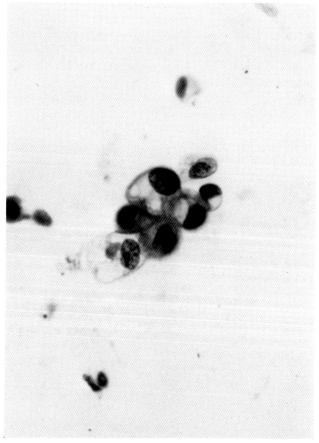

Fig. IX-3. Bronchial washing. Same case as Figure IX-2, demonstrating cells of the adenocarcinoma component of mucoepidermoid carcinoma. Sharply bordered vacuoles are very unusual unless the specimen is received unfixed. Papanicolaou stain, × 600.

evidence of keratinization among the few epidermoid tumor cells. Figure IX-2 does show one typical field of pleomorphic keratinized squamous cancer cells seen with many other cells (Fig. IX-3), demonstrating features of adenocarcinoma, including vacuoles in the cytoplasm. This latter feature is infrequently seen in the authors' cases of bronchogenic adenocarcinoma. The histology of this tumor (Fig. IX-4) is a typical example of mucoepidermoid carcinoma. Included in several fields of this neoplasm were areas of alveolar-type growth pattern.

One author (Frable) has found these cases easy to recognize as pulmonary cancer, but very difficult to diagnose specifically as mucoepidermoid carcinoma. The Department of Cytopathology at Duke University Medical Center has been able to recognize this group of tumors with much more success.[3] Perhaps, in some manner, this is related to the use of fresh sputum preparations, thus producing a greater degree of contrast between vacuolated tumor cells and the epidermoid cells.

Giant-cell carcinomas present no difficulty in diagnosis because they are usually large tumors strategically placed to exfoliate many malignant cells.[4] The basic pattern is that of poorly differentiated adenocarcinoma with the additional feature of very

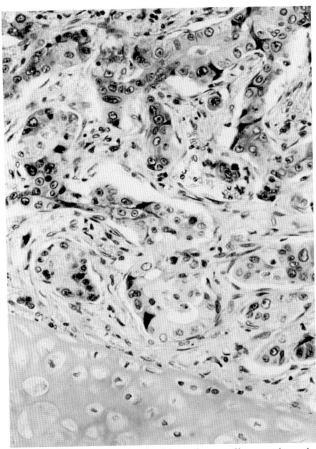

Fig. IX-4. Histologic section of mucoepidermoid carcinoma diagnosed cytologically from the cells in Figures IX-2 and IX-3. Hematoxylin and eosin stain, × 240.

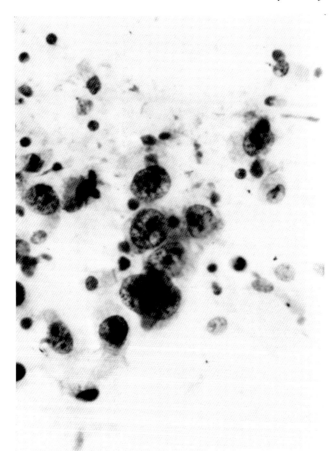

Fig. IX-5. Sputum. Giant-cell carcinoma. Cells demonstrate a pattern of poorly differentiated adenocarcinoma. Note the very prominent nucleoli and degenerated quality of the cytoplasm. Papanicolaou stain, × 600.

pleomorphic giant tumor cells. An example is illustrated in Figures IX-5 and IX-6. The sputum, Figure IX-5, reveals the poorly differentiated adenocarcinoma pattern, the cells having prominent nucleoli, finely vacuolated degenerated cytoplasm, and no well-defined cell borders. Figure IX-6 demonstrates the pleomorphic giant cells, the cell in the upper panel having a histiocytic appearance. An example from a bronchial brushing specimen (Fig. IX-7) depicts pleomorphic and multinucleated tumor cells.

Some giant-cell carcinomas (Fig. IX-8) will be encountered with a basic epidermoid cancer pattern, but will also reveal some very large and pleomorphic tumor cells.[5] Such a lesion is likely to be classified cytologically as a tumor of squamous type unless the evidence of bizarre tumor cells is appreciated. In summary, the cytologic pleomorphism of pulmonary cancer is a good measure of the tissue pleomorphism of the neoplasm.

Rarely, pulmonary cancers become undifferentiated to the point that they assume a spindle-cell pattern of growth. Unless epithelial elements are found in the tissue

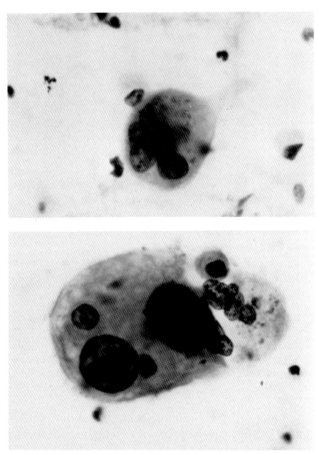

Fig. IX-6. Sputum. Tumor giant cells from giant-cell carcinoma. Same case as Figure IX-5. Note histiocytic appearance of the cell in the upper panel. Papanicolaou stain, × 600.

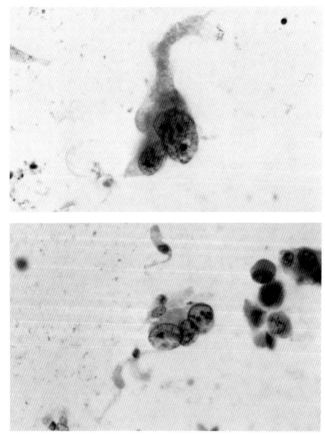

Fig. IX-7. Bronchial brushing. Pleomorphic and multinucleated tumor cells from a case of giant-cell carcinoma. Texture of the cytoplasm is different from that of squamous-cell carcinoma. Papanicolaou stain, × 400.

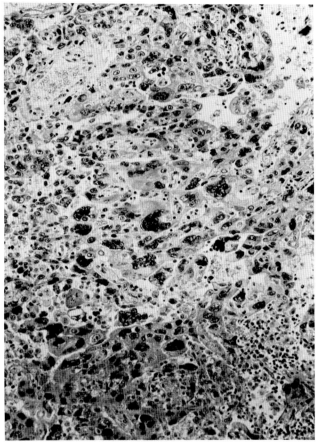

Fig. IX-8. Histologic section. Giant-cell carcinoma diagnosed from cytology depicted in Figures IX-5 and IX-6. Hematoxylin and eosin stain, × 210.

sections, this type of tumor may be considered a true sarcoma. Several cells from such a case are illustrated in Figure IX-9. Malignant characteristics are readily apparent, but the cytoplasmic density would suggest a poorly differentiated squamous-cell carcinoma. The spindle-cell tumor depicted in Figure IX-10 was accompanied by areas with an epithelial pattern of large-cell undifferentiated carcinoma. Spindle-shaped cells can be a prominent feature, actually suggesting a sarcoma rather than a carcinoma. Sarcomas being rather rare in the lung, it is wise to resist the temptation to make that diagnosis from cytology specimens.

Several cases of leiomyosarcomas and four cases of pulmonary blastoma have been reported, as diagnosed from cytologic examination of respiratory specimens.[6-9] Upon review, the cytologic patterns seem rather nonspecific except for the spindle-shaped cells and the density of the cytoplasm in the cases of leiomyosarcoma. Coupled with the clinical presentation, that diagnosis could be suggested. One pulmonary blastoma manifested extremely malignant cells, which favored a large-cell, undifferentiated carcinoma rather than that specific tumor type.[8] In three additional cases the cytologic presentation was that of adenocarcinoma.[9] One of the authors (Frable) has seen one similar case from a thin-needle aspiration biopsy of a chest-wall tumor recurrence following thoracotomy and resection. The cytology was that of an undifferentiated neoplasm without specific morphology. The elongated shape of the tumor cells did suggest a sarcoma more than the combination of epithelial and mesenchymal elements usually seen in pulmonary blastoma.

LYMPHOMAS: HODGKIN'S AND NON-HODGKIN'S TYPE

As expected, there is a dearth of reports of lymphomas diagnosed from respiratory cytology, even though the lung may be a relatively common area of lymphomatous involvement.[10,11] This has been particularly true with modern therapy where the prognosis of the patient has improved and there is additional time for recurrence and dissemination. It may be expected that the cytopathologist will be confronted with more cases in the future, with the increase in longevity of patients with lymphoma and with the relatively easy sampling of the respiratory tract via bronchial brushing.[12] The diagnosis is important in the patient whose immune mechanism may be

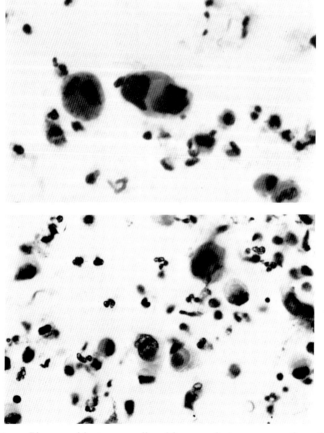

Fig. IX-9. Sputum. Pleomorphic tumor cells with some features suggestive of poorly differentiated squamous-cell carcinoma. Tumor had predominantly a sarcomatous pattern. Papanicolaou stain, × 600.

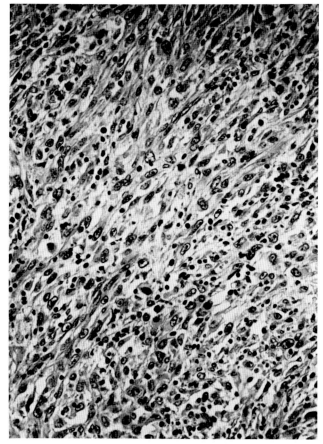

Fig. IX-10. Histologic section demonstrating sarcomatous pattern from the case pictured in Figure IX-9. Other areas showed an undifferentiated large-cell carcinoma. Hematoxylin and eosin stain, × 240.

suppressed by therapy and who then develops a pulmonary infiltrate. Is it a pathologic organism or is it lymphoma?

One author (Frable) has had recent experience with three cases of non-Hodgkin's lymphoma and three cases of Hodgkin's disease encountered as part of an investigation of pulmonary infiltrates in treated lymphoma patients. Two of these cases have been published.[13] All of these cases were seen in the period 1974 through 1977. Only one, a case of Hodgkin's disease primary in mediastinal lymph nodes with invasion of a bronchus, was an untreated primary case of lymphoma at the time of diagnosis. This case is pictured in Figure IX-11. From the initial cytology it was considered an example of large-cell undifferentiated carcinoma. The cells are larger and more irregular than had been found in previous experience with Hodgkin's disease in pulmonary cytology. In retrospect, all of the cells were single tumor cells, a feature that is very unusual for carcinoma. Comparison of the neoplastic cells with the mediastinal lymph node, from which the diagnosis was made, suggests that the two larger cells in Figure IX-11 are Reed-Sternberg cells. A case of secondary Hodgkin's disease of the lung, Figure IX-12, demonstrates the same cytomorphology.

Knowledge of the previous diagnosis is quite important in evaluation of specimens from the respiratory tract with a suspected diagnosis of lymphoma. The cells appear as single cells, although they may lie close to each other in certain areas of the smear (Fig. IX-13). They are usually smaller than the cells of carcinomas, oat-cell tumors excluded. The nuclei may have a very speckled chromatin pattern. There is often a fold or projection of the nucleus (Fig. IX-14) that may reflect differentiation of the lymphoma rather than being a general cytologic characteristic. The tumor cells of lymphoma appearing in an inflammatory background may be very difficult to find. They degenerate in the presence of inflammation, further obscuring the diagnosis. Specimens from bronchial brushing of any infiltrate present may produce cells of much better quality that can be correctly diagnosed.

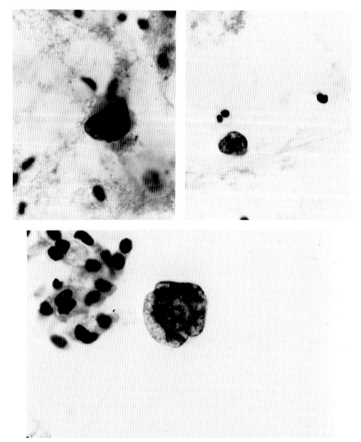

Fig. IX-11. Bronchial washing. Cells of bronchial lymph nodes from a case of Hodgkin's disease invading the respiratory tract. Folds and lobulation of the nuclei are prominent features. Tumor cells all occurred as single cells in this specimen. The case was diagnosed as large-cell undifferentiated carcinoma until compared with a biopsy of a mediastinal lymph node. Papanicolaou stain, × 600.

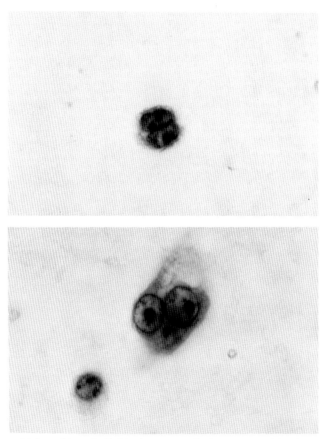

Fig. IX-12. Bronchial washing. Individual tumor cells from case of Hodgkin's disease, with secondary infiltration of the lung. Though smaller, the cell in the upper panel is probably a better example of a Reed-Sternberg cell than the one in the lower panel. Papanicolaou stain, × 1000.

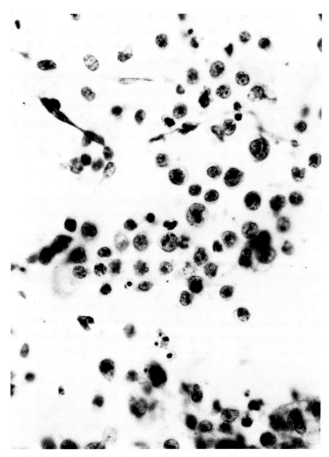

Fig. IX-13. Bronchial washing. Individual cells of poorly differentiated lymphosarcoma found in one area of the smear. Note the very distinct and clumped chromatin pattern. Papanicolaou stain, × 600.

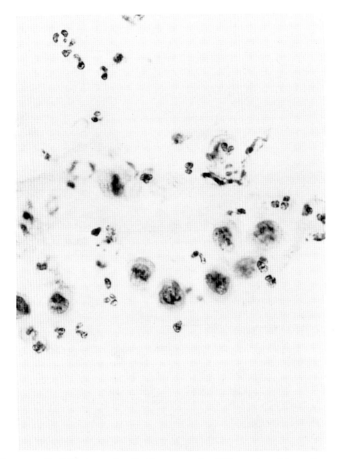

Fig. IX-14. Sputum. Exfoliation of cells from poorly differentiated lymphosarcoma from a patient with an inflammatory-appearing lung infiltrate. Note irregularity of the nuclear membrane and nuclear projections. Papanicolaou stain, × 600.

It is probably not possible to diagnose the so-called lymphocytomas of the lung from cytology. The authors are not aware of any cases of that type in the cytology literature. Theoretically, cells from a lymphocytoma could be obtained with brushing. Presumably they would look like benign lymphocytes having a uniformly hyperchromatic and dense nucleus. Similar cells may be exfoliated spontaneously in sputum, both in relation to inflammatory processes in the lung and from tonsillar tissue of the oropharyngeal area. Experience with cases of lymphocytic lymphoma secondarily involving the lung indicates that, though the cells are small and of uniform size, careful examination of the chromatin pattern reveals that it is open or granular and distinct.

METASTATIC NEOPLASMS

The lung is a common site of metastatic tumors. Although in most cases this means disseminated cancer, treatment may occasionally be successful for a time. The cytopathologist must be concerned with the differential diagnosis of tumor cells from both metastatic lesions and from neoplasms arising in adjacent structures that have invaded some portion of the respiratory tract. The latter may contaminate cytologic specimens. These cases present particular problems, as most of them are squamous-cell cancers of the esophagus and oral cavity. Some examples are demonstrated in Figures IX-15 and IX-16. More of the tumor cells are round and have a dysplastic configuration both in sputum or washings and brushings. This consistent parabasal cell appearance is one feature that suggests contamination by direct extension from an esophageal or oral cancer.

With esophageal carcinoma invading a bronchus, producing a fistula, these round tumor cells are accompanied by an extensive inflammatory infiltrate, both obscuring them and producing degeneration. A brushing specimen will frequently yield better-preserved cells, retaining the same parabasal shape but having obvious nuclear abnormalities indicating malignancy.

Approximately 50% of patients with metastatic tumors to the lung may be expected to exfoliate tumor cells in sputum and/or bronchial washings. The use of brushings, which may sample the lesion directly, can be anticipated to improve this yield. Kern

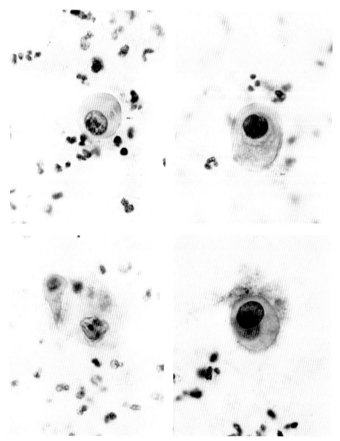

Fig. IX-15. Sputum. Squamous-cell carcinoma of the esophagus with a bronchoesophageal fistula. Cells are dysplastic and primarily parabasal in shape. Many acute inflammatory cells are present in the background. Papanicolaou stain, × 600.

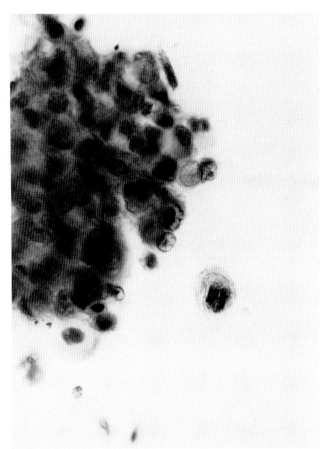

Fig. IX-16. Bronchial brushing. Cluster of round, parabasal-shaped cells of metastatic squamous-cell carcinoma of the larynx to the lung. Papanicolaou stain, × 600.

et al. have reported 71% correct diagnosis of metastatic neoplasms in a summary of their recent experience.[14] They found that most tumors, other than recurrences or metastases from lung cancers, were of the breast.[14] The cytologic patterns of the majority of metastatic neoplasms of the lung are relatively nonspecific. Although some breast cancers present no definite cytologic pattern, many will have the "Indian file" arrangement of tumor cells seen in Figure IX-17, or a large cluster of cancer cells with a smooth community border as depicted in Figure IX-18. Some breast cancers metastatic to the lung bear a striking resemblance to terminal bronchiolo-alveolar cell carcinoma (Fig. IX-19).

Recognizing a specific or unusual pattern in a sputum cytology when definite malignant cells are present may be of some importance clinically. This is quite true if there has been a long interval between treatment of the primary cancer and the appearance of the lung metastases. An additional problem is present if the radiographs suggest a primary cancer of the lung. Such a case is presented in Figures IX-20 and IX-21. X-rays suggested a primary carcinoma of the lung in an elderly woman treated several years previously for a squamous-cell carcinoma of the cervix. Note

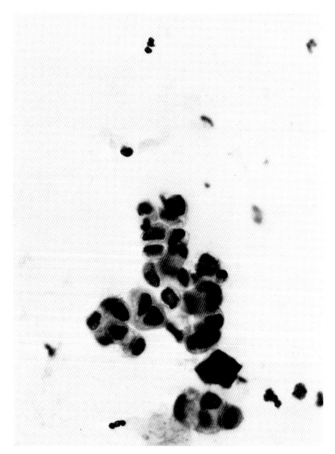

Fig. IX-17. Bronchial washing. Malignant tumor cells demonstrating typical "Indian file" pattern seen in many cases of metastatic breast carcinoma to the lung. Nuclei are very hyperchromatic. Papanicolaou stain, × 600.

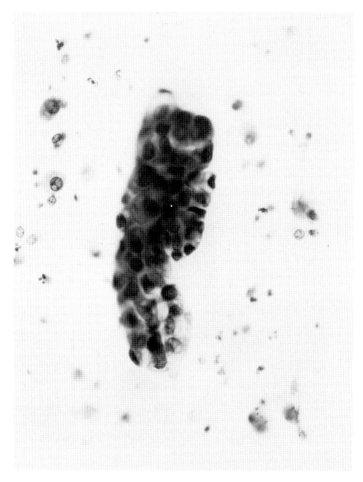

Fig. IX-18. Sputum. Fresh preparation of sputum from a patient with metastatic breast carcinoma to the lung. Note the very large size of the cluster, with smooth community border. Note also the large vacuoles in some of the cells. The cluster is similar to the large cancer-cell balls that are seen in fluids from patients with metastatic breast cancer. The cluster also has some features of a bronchial hyperplastic cell group (Creola body). Papanicolaou stain, × 400.

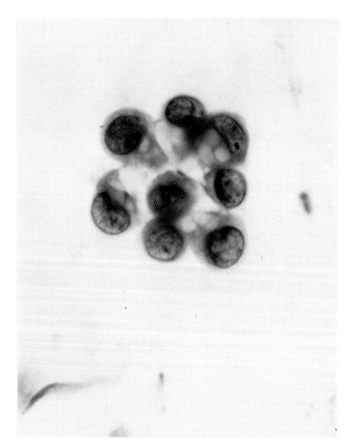

Fig. IX-19. Sputum. Fresh preparation of sputum from another patient with metastatic breast cancer, demonstrating a different pattern. Note the resemblance to the cells from cases of terminal bronchiolo-alveolar cell carcinoma. Compare with Figure VIII-5. Papanicolaou stain, × 1000.

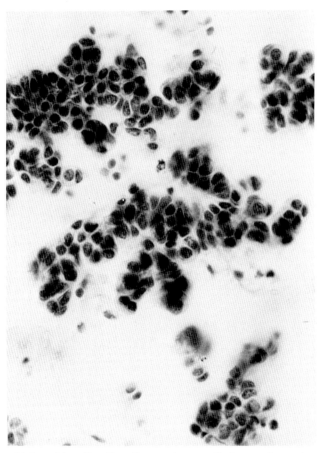

Fig. IX-20. Bronchial washing. Small malignant tumor cells in peculiar trabecular and linear arrangement unlike any primary lung carcinoma. Case of metastatic squamous-cell carcinoma of the cervix to a secondary bronchus, presenting as suspected primary carcinoma of the lung. Papanicolaou stain, × 600.

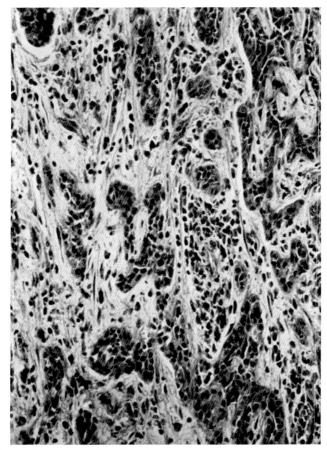

Fig. IX-21. Histologic section of the original cervical biopsy revealing the same trabecular and linear pattern of squamous-cell carcinoma found in the cytology (Fig. IX-20). Hematoxylin and eosin stain, × 240.

the unusual trabecular arrangement of the cells in Figure IX-20. They are the size of cells from oat-cell carcinoma, but the cohesiveness and peculiar linear pattern is very different from that neoplasm. In fact, the cells do not correspond to any cytologic pattern of primary lung cancer. A review of the biopsies from the original cervix primary (Fig. IX-21) demonstrates the same trabecular and linear arrangement of the neoplastic cells in that cancer. A cytologic diagnosis of metastatic squamous-cell carcinoma from the cervix was made and later confirmed by bronchial biopsy and at autopsy.

Although most metastatic tumors of the lung occur as multiple nodules or as a single peripheral mass, some occur as an isolated lesion invading a major bronchus. Exfoliation of cells may be very heavy in this type of case and may be associated with a marked tumor diathesis. This pattern would then suggest a primary lung cancer. Large single and pleomorphic tumor cells are seen in Figure IX-22, from a

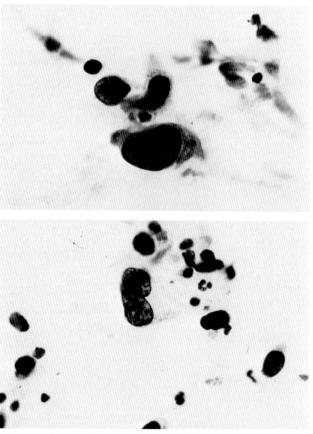

Fig. IX-22. Bronchial washing. Abundant large and irregular tumor cells from metastatic colonic carcinoma to a major bronchus. This case was initially diagnosed as giant-cell carcinoma of the lung until complete history was determined and cytology and bronchial biopsy were compared with original colon primary carcinoma removed six years previously. Papanicolaou stain, × 600.

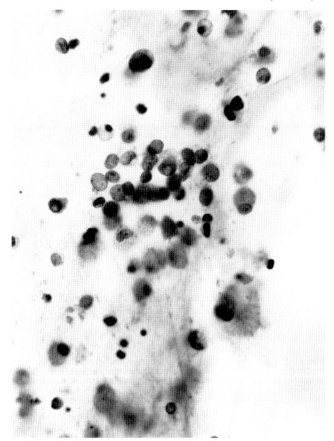

Fig. IX-23. Sputum. Patient with metastatic choriocarcinoma of the testis to the lung. Tumor cells are very undifferentiated and appear relatively small in comparison to the primary tumor cells of choriocarcinomas. There is some resemblance to oat-cell carcinoma, but the age of patients with that tumor usually preclude this diagnosis. Papanicolaou stain, × 400.

case of colonic carcinoma that had been resected six years previously and was now metastatic to a major bronchus. Without the history, this case was considered to be a giant-cell carcinoma of the lung. Bronchial biopsy compared with the original colon primary showed a striking similarity, confirming the diagnosis of metastatic tumor.

Figures IX-23 and IX-24 show two additional examples of metastatic tumors with cytologic patterns resembling primary lung cancers. The small cells with somewhat pale chromatin and a small nucleolus (Fig. IX-23) are from metastatic choriocarcinoma. The general pattern of these cells suggests a possible oat-cell carcinoma. Figure IX-24 shows another example of a cluster of highly vacuolated cells in a fresh sputum from metastatic carcinoma of the colon to the lung. These cells suggest a terminal bronchiolo-alveolar cell carcinoma. Such cases emphasize the importance of providing proper clinical history and present findings to the cytopathologist, particularly if there is a differential diagnosis between primary and metastatic

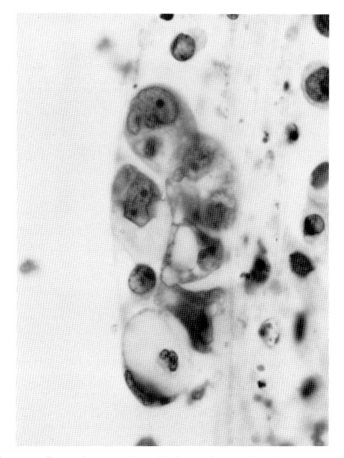

Fig. IX-24. Sputum. Case of metastatic colonic carcinoma. Fresh preparation of sputum demonstrates well-defined vacuoles within these tumor cells. There is a similarity to terminal bronchiolo-alveolar cell carcinoma. Note the very necrotic and inflammatory background characteristic of colonic carcinoma in metastatic sites. Papanicolaou stain, × 1000.

carcinoma. A comparison may then be made between previous tissue sections and the present cytologic specimens, thus enhancing accuracy in correctly differentiating metastatic from primary lung neoplasms.

REFERENCES

1. Auerbach, O., Garfinkel, L., and Parks, V. R.: Histologic type of lung cancer in relation to smoking habits, year of diagnosis, and sites of metastases. *Chest, 67:* 382–387, 1975.
2. Hinson, K. F. W., Miller, A. B., and Tall, R.: An assessment of the World Health Organization classification of the histologic typing of lung tumors applied to biopsy and resected material. *Cancer, 35:* 399–405, 1975.
3. Johnston, W. W. and Frable, W. J.: The cytopathology of the respiratory tract. A review. *Am J Pathol, 84:* 372–414, 1976.
4. Broderick, P. A., Corvese, N. L., LaChance, T., and Allard, J.: Giant cell carcinoma of the lung. A cytologic evaluation. *Acta Cytol, 19:* 225–230, 1975.
5. Oyasu, R., Battifora, H. A., Buckingham, W. B., and Hidvegi, D.: Metaplastic squamous cell carcinoma of bronchus simulating giant cell tumor of bone. *Cancer, 39:* 1119–1128, 1977.

6. Fleming, W. H. and Jove, D. F.: Primary leiomyosarcoma of the lung with positive sputum cytology. *Acta Cytol, 19:* 14–20, 1975.

7. Krumerman, M. S.: Leiomyosarcoma of the lung. Primary cytodiagnosis in two consecutive cases. *Acta Cytol, 21:* 103–108, 1977.

8. Non, D. P., Lang, W. R., Patchefsky, A., and Takeda, M.: Pulmonary blastoma. Cytopathologic and histopathologic findings. *Acta Cytol, 20:* 381–386, 1976.

9. Spahr, J., Draffin, R. M., and Johnston, W. W.: Cytopathologic findings in pulmonary blastoma. *Acta Cytol 23:* 454–459, 1979.

10. Eisenberg, R. S. and Dunton, B. L.: Hodgkin's disease first suggested by sputum cytology. *Chest, 65:* 218–219, 1974.

11. Suprun, H. and Koss, L. G.: The cytological study of sputum and bronchial washings in Hodgkin's disease with pulmonary involvement. *Cancer, 17:* 674–680, 1964.

12. Forrest, J. V. and Sagel, S. S.: Pulmonary infiltrates in lymphoma: value of bronchial brushing. Case reports. *Mo Med, 72:* 240–243, 1975.

13. Smith, R. H., Muren, O., Scott, R. B., and Frable, W. J.: Transbronchial brush biopsy: Ante-mortem diagnosis of pulmonary infiltrates in non-Hodgkin's lymphoma. *Va Med Mon, 105:* 359–362, 1978.

14. Kern, W. H. and Schweizer, C.: Sputum cytology of metastatic carcinoma of the lung. *Acta Cytol, 20:* 514–520, 1976.

CHAPTER

X

Developmental Carcinoma of the Lung: Early Diagnosis

Increased survival for lung cancer patients would seem to rest with early diagnosis.[1] A more logical solution, one that is apparently economically unsound, would be to stop the production and sale of cigarettes. The evidence linking lung cancer to smoking is substantial and persuasive.[2-6] Although this evidence has been challenged, the challenger has been overwhelmed with rebuttal.[7-11] The studies by Doll have recently been reviewed in detail by expert biostatisticians. They found the relationship between smoking and lung cancer highly significant for all age groups.[12]

Saccomanno and coworkers have documented the sequential development of lung cancer in cytologic studies of uranium miners who smoke (Fig. X-1*).[13, 14] The best correlation of lung cancer incidence is that of the combined carcinogenic effects of smoking and uranium mining. Smoking alone correlates next best, while the correlation for uranium mining in nonsmokers is a distant third. The histologic evidence for the association of smoking and lung cancer, as well as smoking and the precancerous lesions of dysplasia, was established earlier than the cytologic findings by Auerbach et al.[3] Subsequent investigations of smoke inhalation using experimental animals and the direct application to the respiratory tract of carcinogens found in tobacco condensates have confirmed the histologic and cytologic observations in man.[2] The experiments of Schreiber et al. are particularly important in this respect.[15, 16]

Smoking experiments in some animals have produced a significant number of adenomas and adenocarcinomas rather than the traditional squamous-cell tumors most consistently linked with smoking in human *males*.[2] The tobacco forces have seized upon this finding as evidence to discredit smoking as a factor in the cause of lung cancer. The recently documented rising incidence of lung cancer in women, of which a disproportionate number is adenocarcinoma, would seem to make that argument specious.[17]

---

* Figure X-1 is reproduced by permission of the authors: Saccomanno, G, et. al.: Developmental carcinoma of the lung as reflected in exfoliated cells. *Cancer 33*, page 259, figure 1, January 1974 and by permission of the publisher, J. B. Lippincott Co., Philadelphia.

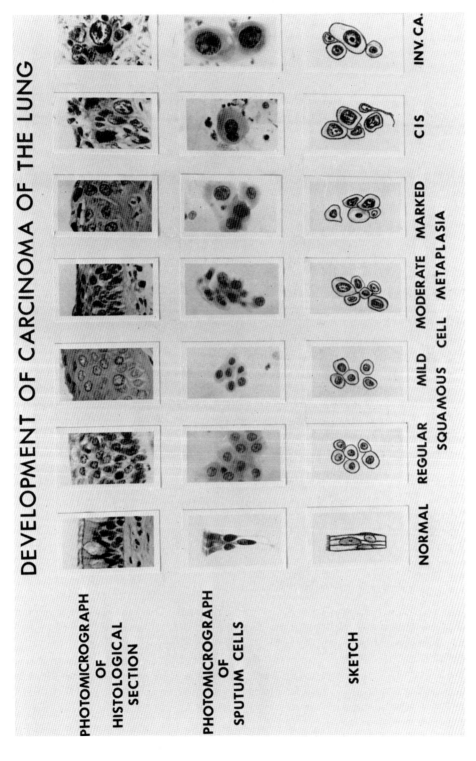

Fig. X-1. Development of carcinoma of the lung. By permission of the authors. Saccomanno G et al: *Cancer 33*: 259, 1974. Figure 1.

The precursor stages of dysplasia and carcinoma *in situ* are thought to exist for an average period of five years respectively, but there is wide individual variation in sensitivity to carcinogenic stimuli.[14, 18-20] Squamous metaplasia as a precursor is frequently found but is an equally common response of the bronchial epithelium to nonneoplastic stimuli.[21-23] In fact, Melamed has denied that squamous metaplasia is a significant precursor of dysplasia and/or carcinoma in situ.[24] His evidence for this seems artificial. He presents a variety of pathways for the development of lung cancer from normal columnar respiratory epithelium and reserve cells, but his diagram, in the authors' opinion, is incomplete.[24]

The sequence of histologic and cytologic changes preceding the onset of other types of lung cancer are much less clear. There is a particular interest in the reserve cell developing through hyperplasia into tumorlets, carcinoid tumors, and oat-cell carcinoma.[25, 26] Although neurosecretory granules have been identified in cells from some of these reserve-cell hyperplasias and tumors, the actual transformation of one to the other has yet to be convincingly demonstrated.[26] This is particularly true of the carcinoid tumors.[25] This tumor may become aggressive, but its clinical behavior even in that form is not that of typical oat-cell carcinoma. The cells in both carcinoid and oat-cell carcinoma are capable of production of precursor amines with hormonal activity, as witnessed by the occasional occurrence of the carcinoid syndrome and the effects of ACTH production seen with some cases of small-cell carcinoma. Recently, proliferation of the reserve cells has been demonstrated experimentally in the hamster following the instillation of carcinogens into the respiratory tract.[27]

Little investigative work has been directed at the developmental stages of adenocarcinoma, perhaps because this tumor has been relatively uncommon until recently. Terikota has induced adenomatosis in the lungs of guinea pigs by producing fibrosis through injection of immune complexes and exposure to high oxygen tension.[28] A morphologically identical reaction can be seen in human lungs under similar insults. Proliferation of alveolar cells and terminal bronchiolar cells has been noted frequently in scarred areas of the lung.[29, 30] No direct connection, however, has been found between these proliferations and the development of terminal bronchiolo-alveolar cell carcinomas. Moreover, the antecedent lesions of bronchogenic papillary and bronchogenic acinar carcinoma are completely unknown.

One of the authors (Frable) has had little personal experience with the cytology of occult *in situ* carcinoma of the lung. Several cases have been diagnosed, but only in retrospect. A patient previously considered to have received a false-positive diagnosis for lung cancer returned several months or years later with radiographic and clinical evidence of carcinoma. One case is illustrated in Figure X-2. The cells seen in the first sputum received were either round or elongated. The nucleus had a highly granular chromatin. The cytoplasm stained lightly eosinophilic or faintly basophilic. The cells in the upper panel of Figure X-2, from the first sputum, were essentially dysplastic, but not different from those seen in response to some inflammatory conditions of the lung. The elongated cells in the lower panel of Figure X-2 had a high nuclear cytoplasmic ratio and distinctly granular chromatin. Except for their shape, they were reminiscent of cells from cases of *in situ* carcinoma of the cervix. Such cells rarely occur in syncytia in sputum. This type of undifferentiated cell seems to have been ignored by those investigators describing the developmental phases of

Fig. X-2. Sputum. Dysplastic cells (upper panels) and a few undifferentiated cells (lower panel) from a patient with occult squamous-cell carcinoma of the lung. Note the abnormal nuclear chromatin and very high nuclear cytoplasmic ratio of the cells in the lower panel. These cells are similar to the cells of invasive poorly differentiated squamous-cell carcinoma, except they are smaller. Compare with Figures V-8 and V-10. Papanicolaou stain, × 600.

squamous-cell carcinoma of the lung.[20, 32, 39] These cells may be preceded by meta-plastic and dysplastic squamous cells or they may occur concurrently. In the authors' cases and those reviewed from the literature, the presence of these cells of undiffer-entiated morphology signals that intramucosal squamous-cell carcinoma at the least is present. Microscopic or even frank invasive carcinoma may have already evolved as well.

In the case described above, the patient began to have symptoms of more frequent cough several months later. Figures X-3 and X-4 demonstrate more of these undif-ferentiated cells. Figure X-3 depicts only nuclear changes of distinct chromatin with very faint cytoplasmic borders to the cells. This almost syncytial arrangement is seldom seen in sputum, but is more frequently found in bronchial washings and brushings. Figure X-4 demonstrates a greater degree of nuclear abnormality, but at the same time there is differentiation of some cells toward the squamous type. These

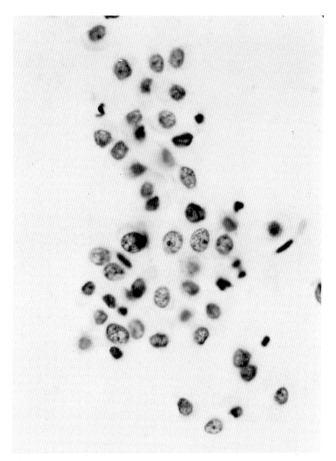

Fig. X-3. Sputum. Same case as Figure X-2. Cells with distinct nuclear chromatin and blurred or absent cytoplasmic boundaries are present. This syncytial pattern is rarely seen in sputum from cases of occult carcinoma of the lung. It may be seen in both washing and brushing specimens when the lesion is being localized. Papanicolaou stain, × 600.

Fig. X-4. Sputum. Same case as Figures X-2 and X-3. Mixture of dysplastic squamous cells and a few undifferentiated nuclei without visible cytoplasm (center of the figure). The dysplastic cells are miniatures of tumor cells found in keratinizing squamous-cell carcinoma. Papanicolaou stain, × 600.

cells may indicate that there is beginning invasion, but they may also only represent areas of mucosal dysplasia, depending upon the stage of evolution of the neoplastic alteration of the bronchial epithelium. Figure X-5 reveals an area of intramucosal squamous-cell carcinoma with a focus of microinvasion. Biopsies were obtained from a slightly roughened and red mucosa from the right lower lobe bronchus. This lesion had probably been present for several years, based upon the current knowledge of the developmental stages of squamous-cell carcinoma of the lung.

A second case is depicted in Figures X-6 through X-8. The cells occurred as single cells with dense and opaque nuclei having an irregular shape. The cytoplasm appeared finely granular, and this, coupled with the appearance of the nuclei, might suggest degenerated histiocytes. Close examination of many of the cells revealed an irregular nuclear border. There was an occasional cell with a dense and orangeophilic cytoplasm, emphasizing that probably all of the atypical cells were of squamous type. This patient did have symptoms of cough and radiographically was found to have a

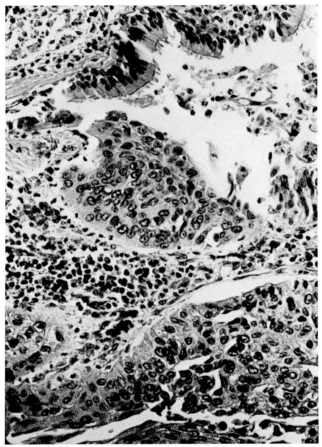

Fig. X-5. Histologic section of biopsy from slightly roughened and red area of mucosa of the right lower-lobe bronchus. A focus of intramucosal squamous-cell carcinoma is seen, with a tongue of microinvasion beneath that area. Hematoxylin and eosin stain, × 240.

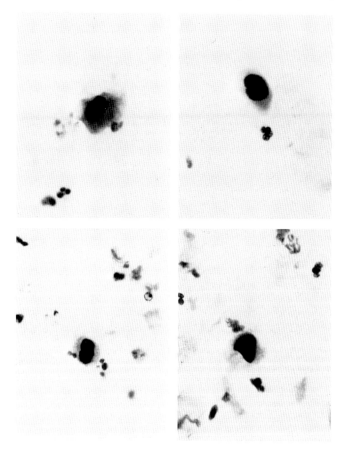

Fig. X-6. Sputum. Small cells with opaque nuclei and granular cytoplasm from a case of occult squamous-cell carcinoma of the lung. These cells might have been considered degenerated histiocytes, except that there were small numbers of dysplastic squamous cells present. Similar cells had been found in several previous sputums about a year before the specimen pictured. Papanicolaou stain, × 600.

small area of pneumonic infiltrate close to the periphery of the left lower lobe. Bronchoscopy was negative; however, segmental bronchography in the area of the pneumonia did demonstrate a focus of slight bronchial obstruction. Without further localization to prove the existence of tumor, a segmental resection of the lung was performed (a plan of management that is not recommended). A small area of obstructing tumor was found in a subsegmental bronchus. Much of the lesion was intramucosal (Fig. X-7), with cells having a very similar morphology to many of those that exfoliated in the sputum. Areas of microscopic invasion were also found (Fig. X-8), originating from a much more differentiated squamous epithelium and showing a differentiated pattern of squamous-cell carcinoma in the areas of invasion. This lesion likewise had probably been present for several years in its incipient stages.

A third and recent case also is another example of a fortuitous diagnosis in a patient under close observation because of a previous resection of a carcinoma of the kidney. Two years after removal of that tumor a routine chest film demonstrated a

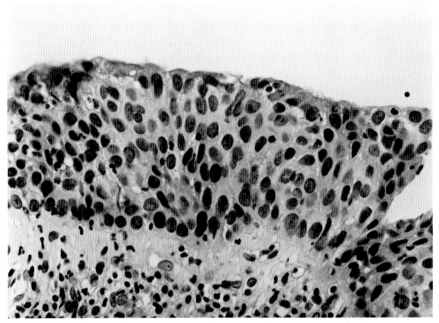

Fig. X-7. Histologic section from segmental resection of left lower lobe illustrates an area of intramucosal squamous-cell carcinoma. Note similarity of the nuclei in this epithelium to those in the sputum in Figure X-6. The resemblance of this intramucosal neoplasm to that seen in intramucosal squamous-cell carcinoma of the cervix is striking. Hematoxylin and eosin stain, × 375.

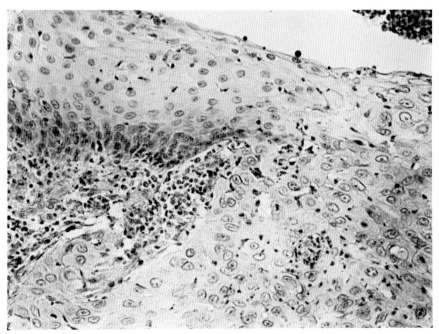

Fig. X-8. Histologic section of area adjacent to that depicted in Figure X-7 demonstrates well-differentiated squamous-cell carcinoma with microinvasion. Much of the surface epithelium of this portion of the neoplasm appears metaplastic. Hematoxylin and eosin stain, × 250.

very slight pneumonic type of infiltrate in the right lower lobe. The patient was not producing sputum and a single induced sample was normal. Bronchial washings of several areas of the right lung were submitted. The cells pictured in Figure X-9 were found in one of these specimens. They have an almost gland-like appearance, with a slightly granular chromatin, uniform small nucleoli, and an increased nuclear cytoplasmic ratio. The cytoplasm is somewhat more dense than in adenocarcinoma cells, but there are some visible cell borders. Even though no mass was seen, the cells were felt to represent metastatic hypernephroma. Compare them with Figure X-10, cells from a case of metastatic hypernephroma. Ignoring the difference in magnification, notice the dissimilarity of the cytoplasm and the size of the nucleoli. Following this initial bronchoscopy, the infiltrate cleared spontaneously. One month later the patient was again bronchoscoped and washings and brushings obtained from the

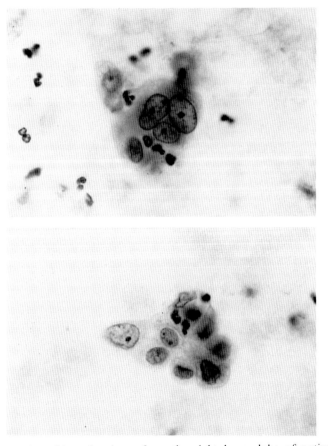

Fig. X-9. Bronchial washing. Specimen from the right lower lobe of patient, with hazy pneumonic infiltrate in that area. The patient had previously had a hypernephroma resected and was under close observation for possible metastatic disease. The cells appear glandular-like, but the cytoplasm is slightly more dense than neoplastic glandular cells, and there are some visible cell borders. The cells were felt to represent metastatic hypernephroma, although no mass lesion was present. The infiltrate of the lung cleared spontaneously. Papanicolaou stain, × 600.

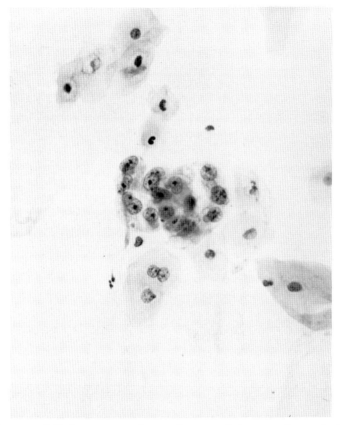

Fig. X-10. Sputum. Cells from an actual case of metastatic hypernephroma to the lung. Note the differences in cytoplasmic texture and more prominent nucleoli in these cells. Compare with Figure X-9. Papanicolaou stain, × 400.

right lower lobe where the pneumonic infiltrate had again appeared. Figure X-11 illustrates cells from this second washing. They are similar to those of the original specimen, also having some squamous features. A few dysplastic and keratinized squamous cells were also found. Cells from the brushing, Figure X-12, are in a sheet-like arrangement and have cell borders. The chromatin, while pale, is granular. The nuclear cytoplasmic ratio is quite high. Associated dysplastic squamous cells were also found in the brushing specimen that was obtained from an area just distal to the junction of the right lower lobe bronchus and the right mainstem bronchus. No visible lesion was seen. With localization from the brushing specimen and the consistent appearance of the infiltrate in the right lower lobe, a thoracotomy was performed and the lobe removed. The bronchi, when opened, revealed no gross lesion. Following fixation, the bronchial tree was step sectioned. One centimeter distal to the line of resection at the bifurcation of the right lateral and posterior basal segmental bronchi the lesion pictured in Figure X-13 was found. It is an intramucosal, partly exophytic squamous-cell carcinoma. A serial section of the entire lesion (an extent of 2.0 mm) did not reveal any areas of invasion. It is the smallest carcinoma of the lung detected by the cytology service of one of the authors (Frable) to date.

The longitudinal studies of Saccomanno and the experimental data from animal models led the National Cancer Institute to support demonstration screening projects among male heavy cigarette smokers 45 years of age or older.[5, 6, 14, 20, 31] These projects, at the Mayo Clinic, Johns Hopkins Hospital, and Memorial Sloan-Kettering Cancer Center are currently nearing completion.[24, 32–34] Although the protocols of the three studies vary, they are all aimed at periodic screening by sputum cytology and radiographic study of heavy cigarette-smoking (presumably high-risk) males. There is a control and a test group; the test group receives a chest x-ray and sputum cytology every four months, while the controls undergo these procedures annually if desired. The Mayo project is in its final phase of screening 10,000 subjects, 5,000 acting as controls. To date the prevalence rate of carcinoma (cases detected at first investigation) has been 0.9%. The incidence rate in the surveillance group is 4.8 cases per 1,000 man-years of observation.[35, 36] Both the prevalence and incidence rates are substantially lower than anticipated for what is considered a high-risk group for developing lung cancer. These rates clearly indicate that it would be futile to screen an unselected population by sputum cytology.

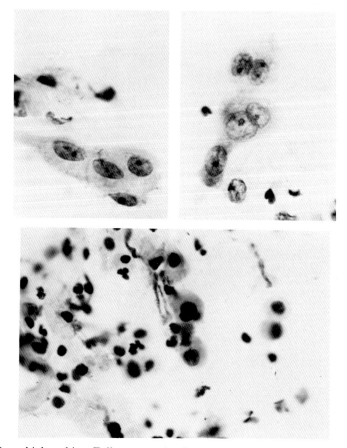

Fig. X-11. Bronchial washing. Follow-up study after reappearance of right lower-lobe infiltrate in patient described in Figure X-9. There are now dysplastic squamous-type cells as pictured in the lower panel. Papanicolaou stain, × 600.

Lesions have been detected in the test group by both x-ray and cytology. The cases found by cytology appear to have a better prognosis than those detected radiographically.[35] These lesions are usually centrally located in major bronchi and their immediate branches. Localization of these central lesions would not have been possible without the development of the flexible fiberoptic bronchoscope and brushing techniques. Each major bronchus, segmental, and subsegmental bronchus, must be brushed and slides prepared. Usually either the bronchial mucosa appears normal or there is only a slight abnormality seen directly with the flexible bronchoscope. This means bronchoscopy is time consuming. Careful attention must be paid to the details of labeling and preparing the slides from each brushing specimen. Once a lesion is localized by cytology, biopsies are also taken before therapy, to confirm both localization and the cytologic diagnosis. As the projects have progressed, therapy has become somewhat more conservative.[35] Documentation of multicentricity of lung

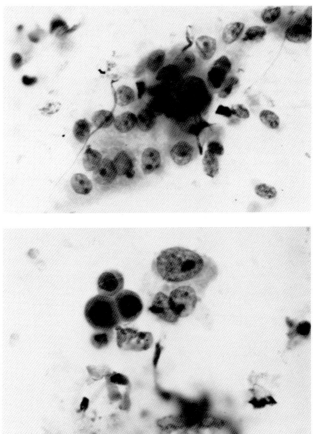

Fig. X-12. Bronchial brushing. Same case as Figures X-9 and X-11. Rather undifferentiated cells are seen in both panels. They have a very glandular-like appearance except for three to the left of center, which show dense cytoplasm and sharp cell borders. A diagnosis of squamous-cell carcinoma was made on both specimens (Figs. X-11 and X-12), with the suggestion that the lesion might be occult, since nothing was seen during bronchoscopy. Papanicolaou stain, × 600.

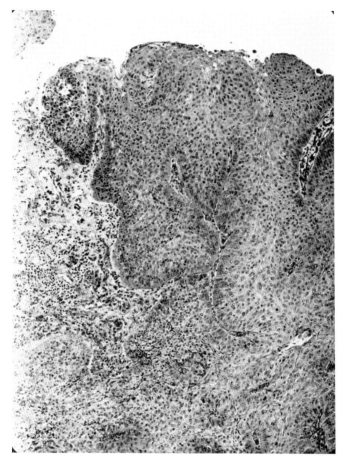

Fig. X-13. Histologic section of lesion detected after step sectioning the junctional area of the right lateral and posterior basal segmental bronchi of the right lower lobe. It is an intramucosal and partially exophytic squamous-cell carcinoma, extending 2.0 mm in length over the bronchial mucosa. There is extension into bronchial glands, but no invasion could be demonstrated. Hematoxylin and eosin stain, × 100.

cancer, not previously a concern because of short arrival, has become important.[37, 38] A few recent cases from the Mayo project have been treated by intrabronchial cryotherapy, since no additional lung tissue could be removed for newly discovered primary cancers.[35, 39]

It was expected that survival of patients in the screening projects would improve by about 60%. Results to date do not approach that anticipated gain in survival. Actuarial calculation of survival at five years for the surveillance group in the Mayo Clinic project is 40%. Of interest is the fact that the control group showed an increase in five-year survival to 25%.[39] Therefore, the net gain from screening by this ambitious method is only 15%. Considering the costs, the authors must question whether screening for lung cancer even in this selected population is worthwhile.

As the incidence of lung cancer shifts in respect to the various histologic types, with an increase already apparent for adenocarcinoma and undifferentiated types, screening any selected population should become even more difficult. Even in high-risk groups, our lack of knowledge of the developmental phase of nonsquamous-cell carcinomas would seem to preclude meaningful results from such a program.

REFERENCES

1. Cohen, M. H.: Guest editorial. Lung Cancer: A status report. *J Natl Cancer Inst, 55:* 505–511, 1975.
2. Auerbach, O., Hammond, E. C., Kirman, D., Garfinkel, L., and Stout, A. P.: Histological changes in bronchial tubes of cigarette-smoking dogs. *Cancer 20:* 2055–2066, 1967.
3. Auerbach, O., Hammond, E. C., Kirman, D., Garfinkle, L., and Stout, A. P.: Changes in bronchial epithelium in relation to cigarette smoking and in relation to lung cancer. *N Engl J Med 265:* 253–267, 1961.
4. Doll, R. and Hill, A. B.: Mortality in relation to smoking: Ten year's observations of British doctors. *Br Med J 5395:* 1399–1410, 1460–1467, 1964.
5. Berlin, N. I.: Early detection and localization of bronchogenic carcinoma. Editorial. *Chest 67:* 508–509, 1975.
6. Berlin, N. I.: Summary and recommendations of the workshop on lung cancer. *Cancer 33:* 1744–1746, 1974.
7. Bross, I. D. J.: Commentary on editorial. The cigarette smoking lung cancer hypothesis. *Am J Public Health 66:* 161, 1976.
8. Higgins, I. T. T.: Commentary, smoking and cancer. *Am J Public Health 66:* 159–161, 1976.
9. Ibrahim, M. A.: Editorial. The cigarette smoking lung cancer hypothesis. *Am J Public Health 66:* 132–133, 1976.
10. Sterling, T. D.: Additional comments on the critical assessment of the evidence bearing on smoking as the cause of lung cancer. Commentary. *Am J Public Health 66:* 161–163, 1976.
11. Sterling, T. D.: A critical reassessment of the evidence bearing on smoking as the cause of lung cancer. *Am J Public Health 65:* 939–953, 1975.
12. Whittemore, A., and Altshuler, B.: Lung cancer incidence in cigarette smokers: Further analysis of Doll and Hill's data for British physicians. *Biometrics 32:* 805–816, 1976.
13. Saccomanno, G., Archer, V. E., Saunders, R. P., Auerbach, O., and Klein, M. G.: Early indices of cancer risk among uranium miners with reference to modifying factors. *Ann N. Y. Acad. Sci. 271:* 377–383, 1976.
14. Saccomanno, G., Archer, V. E., Saunders, R. P., Auerbach, O., and Klein, M. G.: Development of carcinoma of the lung as reflected in exfoliated cells. *Cancer 33:* 256–270, 1074.
15. Schreiber, H., Saccomanno, G., Martin, D. H., and Brenna, L.: Sequential cytological changes during development of respiratory tract tumors induced in hamsters by Benzo(A) Pyrene-ferric oxide. *Cancer Res 34:* 689–698, 1974.
16. Schreiber, H., Saccomanno, G., Martin, D. H., and Brenna, L.: Exfoliative cytology during experimental respiratory carcinogenesis. *Proc Am Assoc Cancer Res 13:* 32, 1972.
17. Miller, A. B.: Recent trends in lung cancer mortality in Canada. *Can Med Assoc J 116:* 28–30, 1977.
18. Saccomanno, G., Saunders, R. P., Klein, M. G., Archer, V. E., and Brennan, L.: Cytology of the lung in reference to irritant, individual sensitivity and healing. *Acta Cytol 14:* 377–381, 1970.
19. U.S. Department of Health, Education, and Welfare, Public Health Service: *The Health Consequences of Smoking, 1974.* DHEW Publication No. (CDC) 74-8704. Superintendent of Documents, Washington, D.C., 1974.
20. Saccomanno, G., Saunders, R. P., Archer, V. E., Auerbach, O., Kuschner, M., and Beckler, P.: Cancer of the lung. The cytology of sputum prior to the development of carcinoma. *Acta Cytol 9:* 413–423, 1965.
21. Carroll, R.: Changes of the bronchial epithelium in primary lung cancer. *Br J Cancer 15:* 215–219, 1961.
22. Kierstenbaum, A. L.: Bronchial metaplasia. Observations on its histology and cytology. *Acta Cytol 9:* 365–371, 1965.
23. Valentine, E. H.: Squamous metaplasia of the bronchus. A study of metaplastic changes occurring in epithelium of the major bronchi in cancerous and non-cancerous cases. *Cancer 10:* 272–279, 1957.
24. Melamed, M. R., Zaman, M. B., Flehinger, B. J., and Martin, N.: Radiologically occult *in situ* and incipient invasive epidermoid lung cancer. *Am J Surg Pathol 1:* 5–16, 1977.
25. Churg, A. and Warnock, M. L.: Pulmonary tumorlet. A form of peripheral carcinoid. *Cancer 37:* 1469–1477, 1976.
26. Churg, A. and Warnock, M. L.: Histopathological studies on tumorlet of the lung with special reference to the cytogenesis of proliferating cells. *Acta Pathol Jpn 25:* 539–553, 1975.

27. Reznik-Schi ler, H.: Proliferation of endocrine (APUD-type) cells during early N-diethylnitrosamine-induced lung carcinogenesis in hamsters. *Cancer Lett 1:* 255–258, 1976.
28. Torikata, C., ᵗakeuchi, H., Yamaguchi, H., and Kageyama, K.: Histopathological studies on experimentally indu ed pulmonary adenomatosis in guinea pig lungs. *Acta Pathol Jpn 25:* 555–563, 1975.
29. Meyer, E. C. a ᵈd Liebow, A. A.: Relationship of interstitial pneumonia, honeycombing and atypical epithelial prolif ᵗration to cancer of the lung. *Cancer 18:* 322–351, 1965.
30. William, J. W.: ᴬlveolar metaplasia: its relationship to pulmonary fibrosis in industry and development of lung cancer. *Brit J Cancer 11:* 30–42, 1957.
31. William, J. W.: Pᵣograms and plans of the National Cancer Institute for research and applications of research methods iᵣ diagnostics to the diagnosis of cancer. *Cancer, 33:* 1705–1711, 1974.
32. Carter, D., Marsh, ᴬ. R., Baker, R. R., Erozan, Y. S., and Frost, J. K.: Relationship of morphology to clinical presentation in ten cases of early squamous cell carcinoma of the lung. *Cancer 37:* 1389–1396, 1976.
33. Marsh, B. R., Frost, ᴶ. K., Erozan, Y. S., and Carter, D.: New horizons in lung cancer diagnosis. *Cancer 37:* 437–439, 1976.
34. Taylor, W. F. and Fontana, R. S.: Biometric design of the Mayo lung project for early detection and localization of bronchogenic carcinoma. *Cancer 30:* 1344–1347, 1972.
35. Fontana, R. S., Sanderson, D. R., Woolner, L. B., Miller, E., Bernatz, P. E., Payne, W. S., and Taylor, W. F.: The Mayo lung project for early detection and localization of bronchogenic carcinoma. A status report. *Chest 67:* 511–522, 1975.
36. Fontana, R. S., Sanderson, D. R., Woolner, L. B., Miller, E., Bernatz, P. E., Payne, W. S., and Taylor, W. F.: The Mayo lung project. Preliminary report of early cancer detection phase. *Cancer 30:* 1373–1382, 1972.
37. McGrath, E. J., Gall, E. A., and Kessler, D. P.: Bronchogenic carcinoma. a product of multiple sites of origin. *J Thorac Surg 24:* 271–283, 1952.
38. Smith, R. A., Nigam, B. K., and Thompson, J. M.: Second primary lung carcinoma. *Thorax 31:* 507–516, 1976.
39. Woolner, L. B.: Personal communication and tutorial lectures, International Academy of Cytology, University of Chicago Center for Continuing Education, Chicago, Illinois. (Unpublished.)

Needle-Aspiration
Biopsy of the Lung

During the past few years there has been a rising interest in needle-aspiration biopsy of the lung.[1] Tissue-needle biopsy, long condemned or considered dangerous, has given way in the United States to the thin-needle aspiration technique of Dahlgren and Nordenström, popular in Europe for some time.[2] Webb has reviewed the history of needle-aspiration biopsy, and the reader may find his article of interest.[3]

Thoracic surgeons, internists, radiologists, and radiotherapists are now performing needle-aspiration biopsy of lung lesions.[4-6] While there is some variation in technique, the external diameter of the needle, 1.0 mm or less, seems to be the critical factor in achieving good results while avoiding serious or frequent complications. Fluoroscopic control or the use of ultrasound to localize the lesion and direct the needle is being used, particularly for small masses.[7-10] One of the authors (Frable) in close cooperation with a thoracic surgeon, has found a simple technique using tape and paper clips (Fig. XI-1) effective for localizing smaller tumors. The clips are placed on a piece of adhesive tape and applied to the patient's skin over the approximate area of the lesion. A film is then taken, revealing the relationship of the paper clips to the mass. Since the clips leave an impression in the skin after the tape is removed, the correct interspace and site to make the aspiration can easily be determined. Technical details of the actual aspiration, directions for correct preparation of the cell sample, and staining techniques are described in the appendix to Chapter II.

Needle-aspiration biopsy may be indicated in any type of lung or mediastinal infiltrate or mass. It is particularly suitable for peripheral lung tumors or those neoplasms in locations that do not produce diagnostic cells by conventional cytology. One of the clearest indications is confirmation of a malignant thoracic tumor in a patient who is clinically inoperable. Radiation therapy and/or chemotherapy may then be instituted promptly.[11]

The cytology of needle aspirates of lung cancers is similar to brushing cytology except that there are more sheets and clusters of tumor cells. Particular patterns and arrangements of cells may indicate even more clearly in this type of specimen the exact histologic classification of the tumor.[12] If the material is abundant, cell blocks

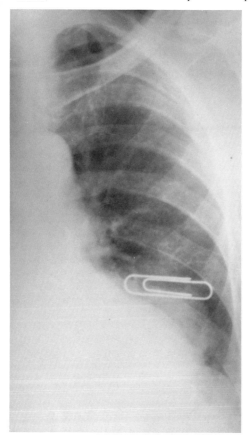

Fig. XI-1. Use of paper clips to determine the correct interspace for aspiration.

may also be prepared, yielding actual tissue fragments for diagnosis. Several cases from various tumor types are illustrated in the figures that follow.

Figures XI-2 and XI-3 demonstrate keratinizing malignant squamous cells aspirated through the supraclavicular space from a superior sulcus tumor in an elderly male patient. A poorly differentiated squamous-cell carcinoma is pictured in Figures XI-4 and XI-5, comparing the air-dried metachrome-B-stained aspiration smear (Fig. XI-4) to the alcohol-fixed, Papanicolaou-stained aspiration smear (Fig. XI-5). The cells from the air-dried smear appear larger, with less nuclear detail. The angular configuration and interdigitation of the cells is highly suggestive of squamous-cell carcinoma. In the Papanicolaou-stained smear, the cell sheets have quite regular and uniform nuclei, with a few small nucleoli visible. Note the higher nuclear cytoplasmic ratio of these cells, as well as the similarity of this sheet from the aspiration smear to cells of poorly differentiated squamous-cell carcinoma found in bronchial brushings (Figs. V-11 and V-12).

Rather than prepare direct smears of needle aspirates, the cytopathologist may prefer filtering the cellular sample. This technique is described in the appendix to Chapter II. Figure XI-6 represents material handled in this fashion from a case of

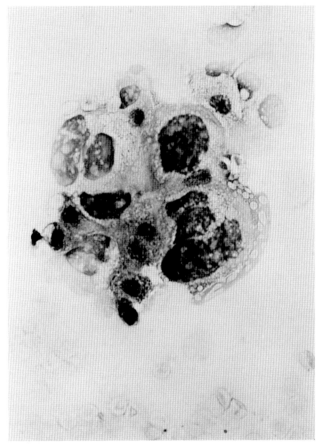

Fig. XI-2. Aspiration. Keratinizing squamous-cell carcinoma. Pleomorphic cells with prominent ectoplasmic border. Degeneration of cells (note vacuolization of nuclei) is commonly found with this tumor. Papanicolaou stain, × 600.

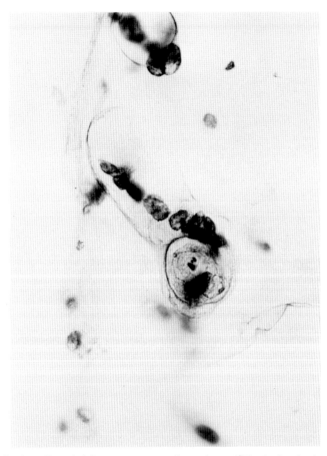

Fig. XI-3. Aspiration. Keratinizing squamous-cell carcinoma. Typical tadpole cell. Papanicolaou stain, × 600.

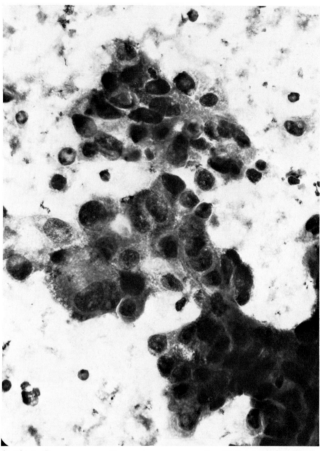

Fig. XI-4. Aspiration. Squamous-cell carcinoma, poorly differentiated. Sheet of cells with round or angular nuclei. Note the interdigitation of the cells as seen in the same tumor in bronchial brushing specimens, Figure V-11. Metachrome B stain, × 500.

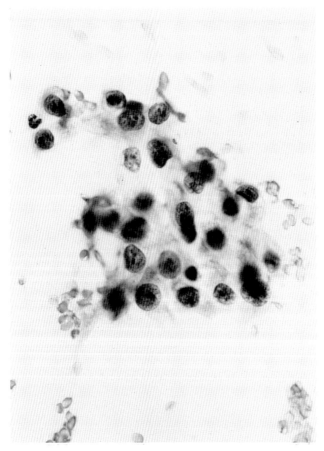

Fig. XI-5. Aspiration. Squamous-cell carcinoma poorly, differentiated. Same case as Figure XI-4. Cells in sheets have well-defined cell borders. Nuclear cytoplasmic ratio is high and there are clearly visible nucleoli. Papanicolaou stain, × 600.

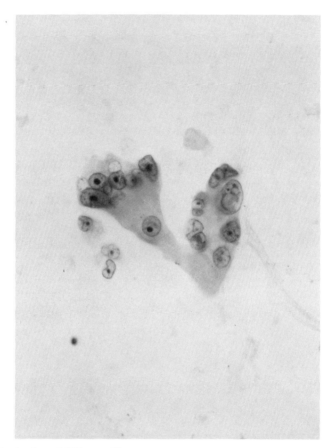

Fig. XI-6. Aspiration, Millipore filter preparation. Large-cell undifferentiated carcinoma. Pleomorphic cells without evidence of cells borders. Note the markedly enlarged nucleoli. Pleomorphism suggests the possibility of giant-cell carcinoma of the lung. Compare with conventional cytology from case of large-cell undifferentiated carcinoma, Figure VI-6. Papanicolaou stain, × 400.

large-cell undifferentiated carcinoma of the lung. Note the prominent nucleoli and absence of cell borders, giving the appearance of a tumor giant cell. There is abundant cytoplasm in comparison to cells of large-cell undifferentiated carcinoma obtained by conventional cytology.

Cases of bronchogenic adenocarcinoma of acinar type will also exhibit sheets of cells in aspiration smears (Fig. XI-7). Like their counterparts in brushing or bronchial washing cytology, the cell borders will be essentially absent, so that the nuclei are actually within a syncytium. Note the very granular chromatin and prominent nucleoli. An additional clue that an aspirate represents adenocarcinoma is the round-to-oval spaces seen within the syncytia. These correspond to the glands formed by the neoplasm. They are consistently present and do not represent fixation artefacts peculiar only to aspiration smears. This pattern of tumor cells may, in some rare instances, be imitated by poorly differentiated squamous-cell carcinoma. Usually,

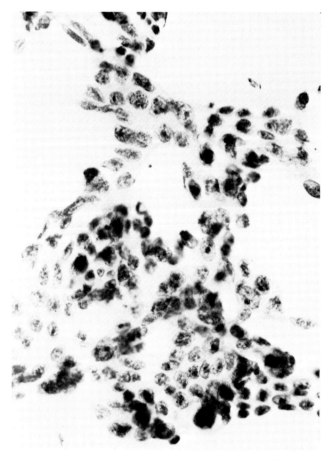

Fig. XI-7. Aspiration. Adenocarcinoma, bronchogenic acinar type. Syncytial arrangement of cells with prominent nucleoli and clumped chromatin. The spaces between cells are not artefacts, but represent glandular spaces seen in tissue sections. Papanicolaou stain, × 375.

however, some cell sheets with good cell borders can be found to provide the necessary diagnostic clue to the correct interpretation of squamous-cell carcinoma.

One of the authors (Frable) has aspirated only one case of terminal bronchiolo-alveolar cell carcinoma, from a female with a pneumonic infiltrate. There were large, three-dimensional clusters of very uniform cells, depicted in Figure XI-8 and XI-9. The cells do not actually look malignant, and small nucleoli that were present are not easily seen in the illustrations. Some of the cells are highly vacuolated, with an almost signet-ring appearance to the nucleus. The depth of focus of the large cell group is impressive, but the cohesiveness is much greater in contrast to sputum and bronchial washing cytology from alveolar cell neoplasms. Individual cells are markedly distorted by neighboring cells, suggesting a bronchial or alveolar cell hyperplasia rather than carcinoma. Pleural-fluid cytology revealed the same cell configurations, confirming the diagnosis of adenocarcinoma. The alveolar pattern of growth of the tumor was present in later follow-up tissue sections.

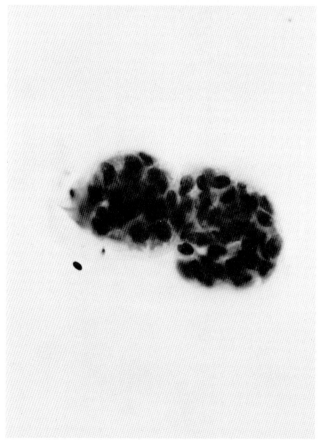

Fig. XI-8. Aspiration. Terminal bronchiolo-alveolar cell carcinoma. Very cohesive cluster of cells with uniformly hyperchromatic nuclei. There is depth of focus, but the community border to the cell group and good nuclear and cell molding suggest a reactive process rather than carcinoma. Papanicolaou stain, × 600.

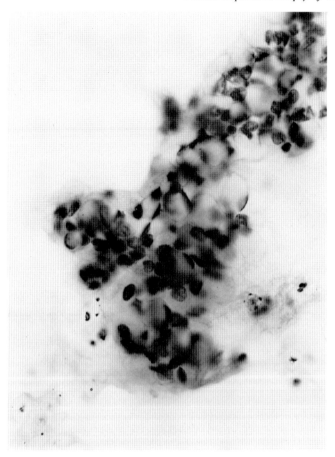

Fig. XI-9. Aspiration. Terminal bronchiolo-alveolar cell carcinoma. Same case as Figure XI-8. The three-dimensional effect of this large cell group is a dominant feature. Note the highly vacuolated cells. The configuration more closely resembles the tissue pattern of alveolar-cell carcinoma in comparison to the cell groups in sputum, bronchial washings, or brushing cytology. Compare with Figures VIII-3 and VIII-6. Papanicolaou stain, × 600.

Small-cell or oat-cell carcinoma in transthoracic aspirates is nearly identical cytologically to the cells seen in brushing specimens. An aspirate from a mid-lung mass is pictured in Figure XI-10. There is a good tumor diathesis about the malignant cells and some evidence of nuclear molding. The clumped chromatin structure and the presence of nucleoli are more easily seen in this type of specimen, perhaps because of better preservation and sampling from the viable areas of tumor. Aspirates of necrotic areas will show degenerated tumor cells found in conventional sputum cytology and large amounts of nucleoprotein that produce the artefact in tissue biopsies familiar to pathologists. A filter preparation from a transthoracic aspiration of oat-cell carcinoma is shown in Figure XI-11 for comparison.

Transthoracic needle-aspiration biopsy is also useful in cases with lung infiltrates in which there is a differential diagnosis between a neoplastic process and an inflammatory one. The cause of the inflammation may not be specifically recogniz-

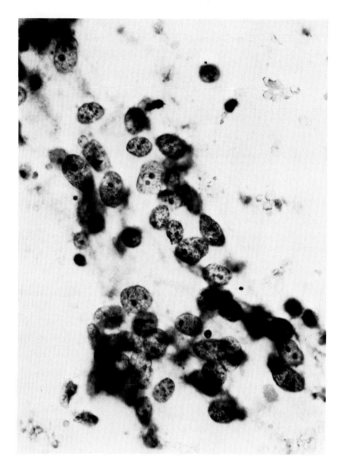

Fig. XI-10. Aspiration. Small-cell undifferentiated carcinoma. Note the similarity in size and appearance of the cells from this carcinoma to those obtained from the same tumor by bronchial brushing. Compare with Figures VII-8 and VII-9. Papanicolaou stain, × 600.

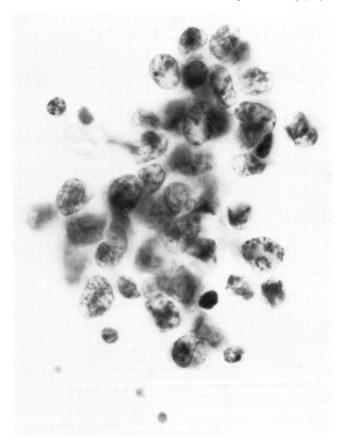

Fig. XI-11. Aspiration, Millipore filter preparation. Small-cell undifferentiated carcinoma. Note the very granular nuclear chromatin. Nucleoli are not as obvious in this example of small-cell carcinoma. Papanicolaou stain, × 1000.

able, but neoplasia can be ruled out. The following case documents two separate transthoracic aspirations in a young boy for the diagnosis of diffuse lung infiltrates (Figs. XI-12 through XI-14).

This child was being treated for acute lymphoblastic leukemia when he presented with dyspnea and radiographic evidence of a hazy, diffuse lung infiltrate. Approximately 10 days before the onset of these symptoms evidence of exacerbation of the leukemia was detected after examination of the bone marrow, and the child was begun on vincristine and daunorubicin. *Pneumocystis carinii* infection was considered a probable etiology for the lung infiltrate. Transthoracic thin-needle aspiration biopsy was carried out without difficulty. No organisms of pneumocystis were identified, but the aspirate did reveal large, reactive alveolar pneumocytes (Fig. XI-13). This type of reactive cell pattern, while not specific, had been seen in other patients treated with combination chemotherapy, including the two drugs vincristine and daunorubicin. Previous patients with this clinical picture had responded to steroids. An excellent response occurred in this patient. He was well and his leukemia remained

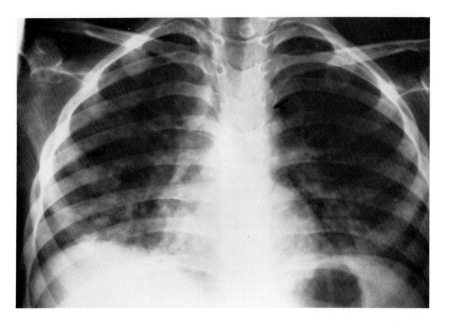

Fig. XI-12. Radiograph of chest of young male being treated with chemotherapy for acute lymphoblastic leukemia. Hazy infiltrate extends peripherally from the hilum bilaterally. Film taken just prior to transthoracic needle-aspiration biopsy.

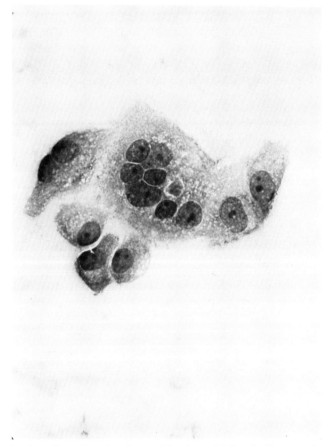

Fig. XI-13. Aspiration. Reactive alveolar and/or bronchial cells from patient described in Figure XI-12. The cause of these reactive cells remains unknown but may be related to combination chemotherapy or perhaps a virus infection. Compare with Figures XI-7 and XI-8 from cases of adenocarcinoma of the lung. Metachrome B stain, × 600.

in remission for another year, at which time he returned with the same symptoms and identical radiographic findings. Transthoracic aspiration biopsy (Fig. XI-14) documented that exacerbation of the leukemia had taken place in the lung. The patient was refractory to further therapy and expired a short time later.

Organisms with an identifiable morphology may be diagnosed by this technique. The reader is referred to the recent article by Bhatt et al.[13] The morphology of organisms detected by this technique is essentially similar to that seen in cytologic specimens of respiratory material. In particular, *Pneumocystis carinii* is unusually amenable to detection by needle-aspiration biopsy. The specific features of the trophozoites as identified with the Giemsa stain are pictured in Figure XI-15.

Another infiltrate of the lung, where specific alterations in the cells confirm the diagnosis, may be observed in Gaucher's disease (Fig. XI-16). This rare case, one of the few reported with an antemortem diagnosis of infiltration of Gaucher's cells in the lung, was diagnosed from a transthoracic needle-aspiration biopsy performed on

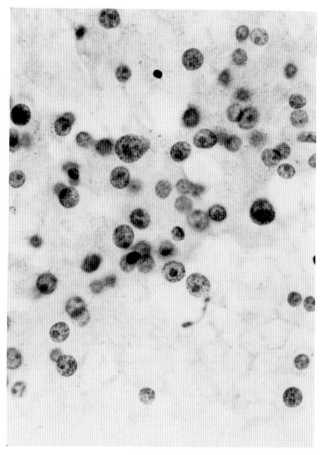

Fig. XI-14. Aspiration. Same case as Figures XI-12 and XI-13. The patient presented one year later with the identical x-ray picture of the chest and dyspnea. Many lymphoblasts in the aspirate reveal that the leukemia has exacerbated in the lung. The patient was refractory to further therapy. Papanicolaou stain, × 600.

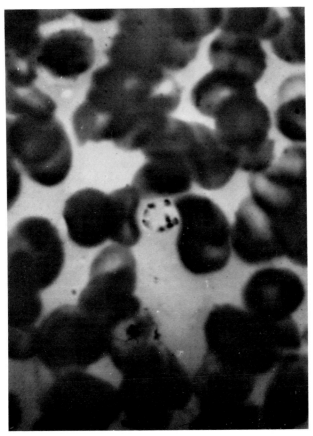

Fig. XI-15. Aspiration. Radial arrangement of trophozoites within a cyst of *Pneumocystis carinii*. Cyst wall stains with methenamine silver, but the actual trophozoites are not seen. May-Grunwald Giemsa stain, × 1000.

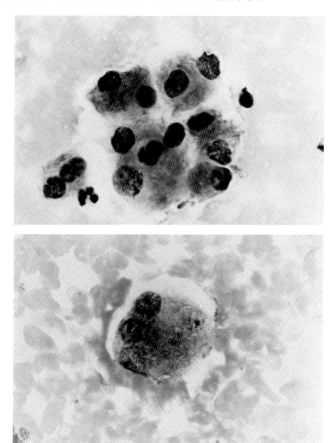

Fig. XI-16. Aspiration. A group of histiocytic cells and a single, multinucleated histiocyte with PAS-positive cytoplasm. Note the irregular cytoplasmic lines characteristic of Gaucher's cells. Gaucher's disease in a two-year-old male with infiltration of the lungs. Periodic acid Schiff (PAS) stain, × 600.

a two-year-old male. Although the patient was known to have Gaucher's disease, the lung infiltrate suggested that an inflammatory process was a more likely diagnosis. Massive infiltration of the lung with Gaucher's cells was confirmed at autopsy (Fig. XI-17).

Accurate interpretation of aspirates from lung lesions requires close cooperation between clinician and cytopathologist. The patterns of cells revealing carcinomas remain constant and should be followed closely. Proliferations of alveolar and bronchial cells in nonspecific reactive and inflammatory conditions may be impressive, leading to an incorrect diagnosis of cancer. Features of regeneration and repair duplicate the same process in other types of respiratory tract cytologic specimens. There are usually many more reactive cells in aspirations and the cell sheets are large. Nuclear cytoplasmic ratios, however, are still normal for the relative immaturity of some of the cells. Inflammatory cells are frequently found within the cytoplasm of the reactive cells, as described previously. Results of the personal experience of one of the authors are summarized in Table XI-1.

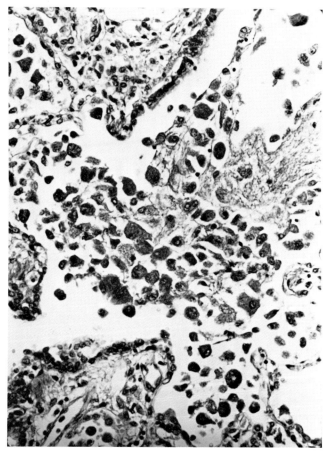

Fig. XI-17. Histologic section of lung with bronchi and alveoli filled with Gaucher's cells. Autopsy of patient whose aspiration biopsy of the lung is illustrated in Figure XI-16. Periodic acid Schiff (PAS) stain, × 240.

The most frequent complication following aspiration biopsy has been pneumothorax. Six percent of the patients in the Medical College of Virginia series developed pneumothorax that required chest-tube drainage as treatment. This is comparable to figures in the literature.[2, 14] It can be expected that all patients who are aspirated while receiving ventilatory assistance will develop pneumothorax, requiring chest-tube drainage. This accounts for half of the cases of this complication in the present experience. Those patients with high respiratory rates will usually not escape pneumothorax, accounting for nearly all the remaining patients in this series. A small amount of air in the pleural space can be demonstrated radiologically in 75% of patients, immediately following transthoracic aspiration biopsy, yet the actual number of patients requiring treatment should be minimal. Symptomatic pneumothorax is usually manifested within one-half hour following the aspiration biopsy. One of the authors (Frable) has had two cases where it was delayed (4 hours and 24 hours). Other complications resulting from aspiration biopsy in this series are a fatal case of air embolism and a transient bout of hemoptysis.

TABLE XI-1

Thin-Needle Aspiration Biopsy of Lung—Results and Complications

Primary malignant tumors	Metastatic tumors	Benign inflammation or other	False positive	False negative	Complications by diagnosis
35	19	25	$(1)^a$	$(5)^b$	None
3	2	4			Pneumothorax^c
		1			Vagal Reflex
		1			Air Embolism
1					?Bleeding
		1			Hemoptysis

92 aspirations in 85 patients

^{a} Carcinoma suspected from aspiration. Diagnosis at thoracotomy organizing pneumonia.

^{b} One case of mesothelioma proven at thoracotomy. Two cases proven metastatic tumors to distant sites, with lung infiltrate the most likely primary site. Two cases of primary lung cancer missed on first attempt at aspiration and diagnosed on second attempt.

^{c} Pneumothorax requiring treatment with tube or aspiration of the air.

Contraindications to aspiration biopsy of the lung have been well described in the literature and they should be observed.[15] They are:

1. Hemorrhagic diathesis.
2. Anticoagulant therapy.
3. Severe pulmonary hypertension.
4. Pulmonary hydatid cyst.
5. Uncontrolled cough.
6. Advanced emphysema.
7. Suspected arterio-venous malformations.

REFERENCES

1. Deeley, T. J.: *Needle Biopsy.* Butterworths, London, 1974, pp 40–50.
2. Dahlgren, S. E. and Nordenström, B.: *Transthoracic Needle Biopsy.* Almquvist & Wiksell, Stockholm, 1966.
3. Webb, A. J.: Through a glass darkly. (The development of needle aspiration biopsy). *Bristol Med Chir J, 89, 322:* 59–68, 1974.
4. Borgeskow, S., and Francis, D.: A comparison between fine-needle biopsy and fiberoptic bronchscopy in patients with lung lesions. *Thorax, 29:* 352–354, 1974.
5. Millard, J. R. and Westcott, J. L.: Percutaneous needle washings in the diagnosis of cavitary lesions of the lung. *Radiology 111:* 474, 1974.
6. Sinner, W. M. and Sandstedt, B.: Small-cell carcinoma of the lung. Cytological, roentgenologic, and clinical findings in a consecutive series diagnosed by fine-needle aspiration biopsy. *Radiology 121:* 269–274, 1976.
7. Chandroasckhar, A. J., Reynes, C. J., and Churchill, R. J.: Ultrasonically guided percutaneous biopsy of peripheral pulmonary masses. *Chest 70:* 627–630, 1976.
8. King, E. B. and Russell, W. M.: Needle aspiration biopsy of the lung—technique and cytologic morphology. *Acta Cytol 11:* 319–324, 1967.
9. Pavy, R. D., Antic, R., and Begley, M.: Percutaneous aspiration biopsy of discrete lung lesions. *Cancer 34:* 2109–2117, 1974.
10. Yam, L. T. and Levine H.: Rapid cytologic diagnosis of percutaneous needle aspirates of peripheral pulmonary lesions. *Am J Clin Pathol 59:* 648–652, 1973.
11. Walls, W. J., Thornbury, J. R., and Naylor, B.: Pulmonary needle aspiration biopsy in the diagnosis of Pancoast tumors. *Radiology, 111:* 99–102, 1974.

12. Sassy-Dobray, G., Keszler, P., and Kompolthy, D.: Experiences with respect to intraoperative cytodiagnosis. *Acta Cytol 16:* 478–482, 1972.
13. Bhatt, O. N., Miller, M. Sc. R., Riche, J., and King, E.: Aspiration biopsy in pulmonary opportunistic infections. *Acta Cytol 21:* 206–209, 1977.
14. Hayata, Y., Oo, K., Ichiba, M., Goya, Y., and Hayashi, T.: Percutaneous pulmonary puncture for cytologic diagnosis—its diagnostic value for small peripheral pulmonary carcinoma. *Acta Cytol 17:* 469–475, 1973.
15. Sanders, D. E., Thompson, D. W., and Pudden, B. J. E.: Percutaneous aspiration lung biopsy. *Can Med Assoc J 104:* 139–142, 1971.

Index

(Page numbers for photomicrographs are in *italics*.)

311